PENGUIN BUSINESS

IMPACTFUL DATA VISUALIZATION

Kavitha Ranganathan is a faculty in the information systems area at the Indian Institute of Management Ahmedabad (IIMA). She holds a master's degree and a PhD in computer science from the University of Chicago and a master's degree in information systems from BITS, Pilani.

Kavitha has been teaching data visualization courses to MBA and doctoral students, analytics professionals and in a wide range of corporate training programmes for more than a decade now. She draws from this vast and rich experience for her book, which captures the main principles for avoiding misleading graphs and creating effective and intuitive visualizations.

Kavitha lives in Ahmedabad with her husband, their two kids and Coco, the youngest furry canine addition to the family. Whenever she gets a free moment, Kavitha can be found walking around the IIMA campus, admiring the many species of trees and birds in this gorgeously green oasis.

INDIA'S BESTSELLING BUSINESS BOOKS SERIES

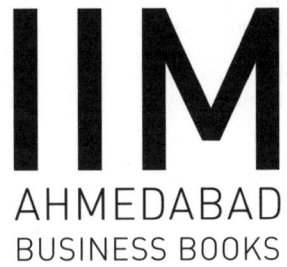

AHMEDABAD
BUSINESS BOOKS

IMPACTFUL DATA VISUALIZATION

Hide and Seek with Graphs

KAVITHA RANGANATHAN

PENGUIN
BUSINESS

An imprint of Penguin Random House

PENGUIN BUSINESS

USA | Canada | UK | Ireland | Australia
New Zealand | India | South Africa | China | Singapore

Penguin Business is part of the Penguin Random House group of companies
whose addresses can be found at global.penguinrandomhouse.com

Published by Penguin Random House India Pvt. Ltd
4th Floor, Capital Tower 1, MG Road,
Gurugram 122 002, Haryana, India

Penguin
Random House
India

First published in Penguin Business by Penguin Random House India 2023

10 9 8 7 6 5 4

ISBN 9780143461746

Typeset in Bembo Std
Printed at Replika Press Pvt. Ltd, India

www.penguin.co.in

MIX
Paper | Supporting
responsible forestry
FSC
www.fsc.org   FSC™ C016779

For Coco, who teaches me
every day to live in
exuberance and love with abandon.

CONTENTS

CHAPTER 1

INTRODUCTION

Let us start with the obvious. We are surrounded by data. We as a generation are neck deep in data and are trying hard to stay afloat in this ocean we ourselves have carefully generated, recorded and are trying hard to leverage. From organizations which strive to be 'data-driven' and students who want to specialize as 'data scientists' to politicians who try to use 'big data' to swing elections and school administrators who collect data to identify effective teachers, the lure of the power of data is widespread. Of course, the crucial link to being able to leverage all the data we diligently collect is to be able to make sense of it or, in other words, convert the data into meaningful information which can be used.

Data visualization is increasingly seen as an effective tool to both make sense of data and communicate insights gleaned from this data to your audience. Traditionally, this audience might have been the CEO of a company listening to a presentation on the company's performance in the last quarter or a physicist reading a journal article about the latest particle physics experimental results. However, today, the audience for data visualization could be almost anyone. It could be a cricket fan trying to understand the batting performance of their favourite player on a sports web portal or a parent trying to decipher their child's board exam results as reported in the marksheet vis-à-vis the national performance. Data visualization (commonly called charts or graphs) works at an intuitive level and has the ability to take a bunch of seemingly meaningless numbers and convert them into a meaningful insight. This in essence is the power of a good graph. A visualization's ability to convey an insight from a collection of numbers is far greater than say a corresponding table of numbers and is universal in its appeal. This, to a large extent, explains why charts and graphs can be found in use in a variety of contexts and scenarios today.

However, data visualization is only a tool, which can be used to spot nuggets of insights from large amounts of data and present these insights in a more palatable form. Typically, this entails constructing a more abstract or, in other words, a higher-level view of the basic information, and therein lies the scope for subterfuge. The transition from plain numbers to a visual form carries with it ample scope to 'massage' the data to convey one's intended message. Data visualization is a powerful but double-edged sword. Creating a misleading graph which conveys a particular idea or narrative which may not necessarily be true is relatively easy as we shall find out soon, especially if the audience is unsuspecting or ill-informed.

One would hope that such instances of misleading graphs would be few and far between. However, as the use of data visualization expands into more and more walks of life, the scope for creating untruthful graphs to mislead unaware audiences only seems to be increasing. It should be noted that not all untruthful graphs are intentionally misleading. Sometimes the creator of the visual themselves may not realize that their graph conveys false information.

For someone like me who spends a large amount of time studying, teaching and practising effective data visualization, every misleading graph in an organization's annual report or a newspaper article is one too many. This book is an attempt to identify and explain different types of misleading data visualizations which are in common use. Drawing from evidence-based research studies, the book offers graph design aspects and alternate constructs which can be used to avoid these deceptive graphs altogether and move towards honest, truthful visualizations.

Why is this topic important enough to warrant an entire book? As someone who has spent more than a decade teaching data visualization to a range of audiences (undergraduates, business school students, top-level corporate executives, analytics professionals and doctorate candidates), what still amazes me is the shock and disbelief when my class participants encounter some of these misleading graphs. Some realize for the first time that a certain graph design is incorrect and untruthful. For many others, it is a big revelation that not only are certain graphs untruthful but they are also in common circulation among us. Most are appalled by the fact that many reputed brands and institutions have at some point used one or more of these ploys. This book is thus my attempt to circulate this knowledge as widely as possible so that anybody dealing with data visualization will be wary of untruthful graphs.

Before proceeding, we should make a clear distinction between an ineffective graph vis-à-vis a misleading one. While the former may just be unsuccessful in bringing out a particular message which is supported by the base data, the latter (either intentionally or unintentionally) conveys a message which is not substantiated by the data. This book covers both these categories and is intended for anyone who creates or consumes any kind of data visualization and would *not* want to be misled by or mislead others. The large number of examples covered will help you instantly recognize a dishonest construct and avoid the 'hide and seek' of data in graphs. Each misleading graph discussed in this book is accompanied by suggestions for alternate designs and constructs which are empirically proven to be more effective. By understanding the art and science behind effective data visualizations, you will be able to communicate the message in your data in a clear, intuitive and accurate manner.

WHETTING YOUR APPETITE

Let's quickly dive into some common examples of what I have been hinting at. Examine the four graphs shown below. Each graph is misleading in its own unique way.

Figure 1.1 is a bar chart which depicts nicotine levels in four popular cigarette brands. Which brand seems the least dangerous? Can you figure out how this graph misleads the viewer? Figure 1.2 contains a pie chart which shows the market share for three different companies. Which company has the largest market share? In what way is this pie chart misrepresenting information?

Figure 1.3 is a line graph which shows profits over time for a particular company. This graph is also misrepresenting data. Can you figure out why it is a misleading graph?

Figure 1.4 is an area chart which represents monthly expenses across four departments in a company. Do all departments show a similar expenditure pattern of peaking in March? If you are tempted to answer 'yes', look again!

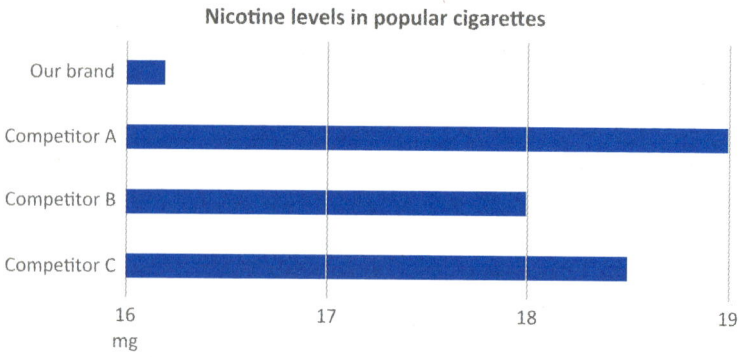

Figure 1.1: Bar chart of nicotine levels in cigarette brands.

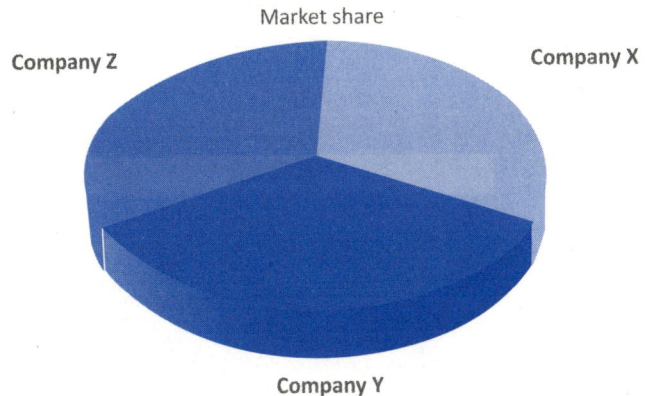

Figure 1.2: Pie chart of market share for three companies.

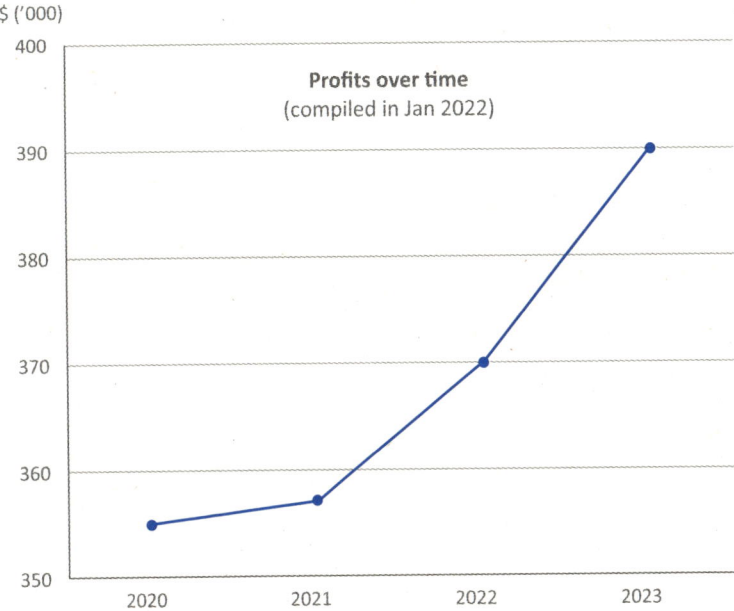

Figure 1.3: Trend line of annual profits.

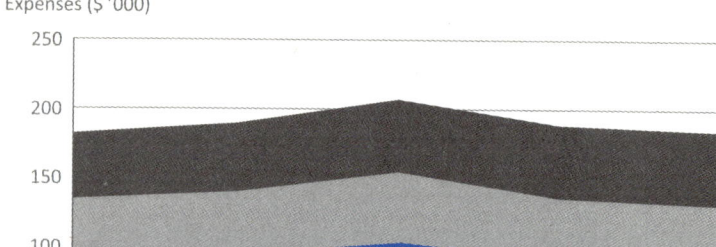

Figure 1.4: Area chart of department-wise monthly expenses.

In the chapters that follow, we will examine each of these graphs (along with many more) to understand how these might mislead the viewer to a false conclusion. For those who cannot wait to know the answers to the four examples shown above, you can jump right away to the relevant sections in the book where they are discussed, by referring to the information below.

- Nicotine level bar chart (Page 25)
- Market share pie chart (Page 51)
- Profits line chart (Page 77)
- Department expenditure area chart (Page 82)

Before we proceed, here is a quick recap of some of the basic chart types you will encounter in the chapters to follow.

BASIC CHART TYPES

While there are dozens of different types of charts which have been invented, many of them are rather esoteric and rarely used.

For this introduction, we will concentrate on the chart types which are commonly seen and most widely used.

If you want to know more about the entire range of chart types which have been created, Severino Ribecca's 'The Data Visualization Catalogue'[1] and Robert Harris's *Information Graphics: A Comprehensive Illustrated Reference* are good sources.

One of the most useful ways to categorize charts is by their function or, in other words, the kind of relationship in the data they are best at highlighting. The high-level functional categorization used here is chiefly influenced by Stephen Few's approach, as elaborated in his seminal book *Show Me the Numbers*.

The six commonly occurring relationships in data elements are comparison/ranking, trends, correlation, distribution, hierarchy and composition. Let's dive into what each of these relationships mean and the common constructs which serve well for each one, as illustrated in Figures 1.5 and 1.6.

Comparison/Ranking
Bar Chart | Column Chart | Lollipop Chart | Slope Graph

When individual values need to be compared with each other and sometimes more specifically in a sorted order, we are dealing with a comparison or ranking relationship, for example, individual departments sorted by their total expenses in the last quarter.

Bar charts (horizontal bars) or **column charts** (vertical bars) work particularly well for such scenarios. Are there any advantages of one over the other? Typically, if a ranking relationship needs to be highlighted, a bar chart might work better than a column chart since we are more accustomed to ranks which flow from

---

[1] http://www.daavizcatalogue.com

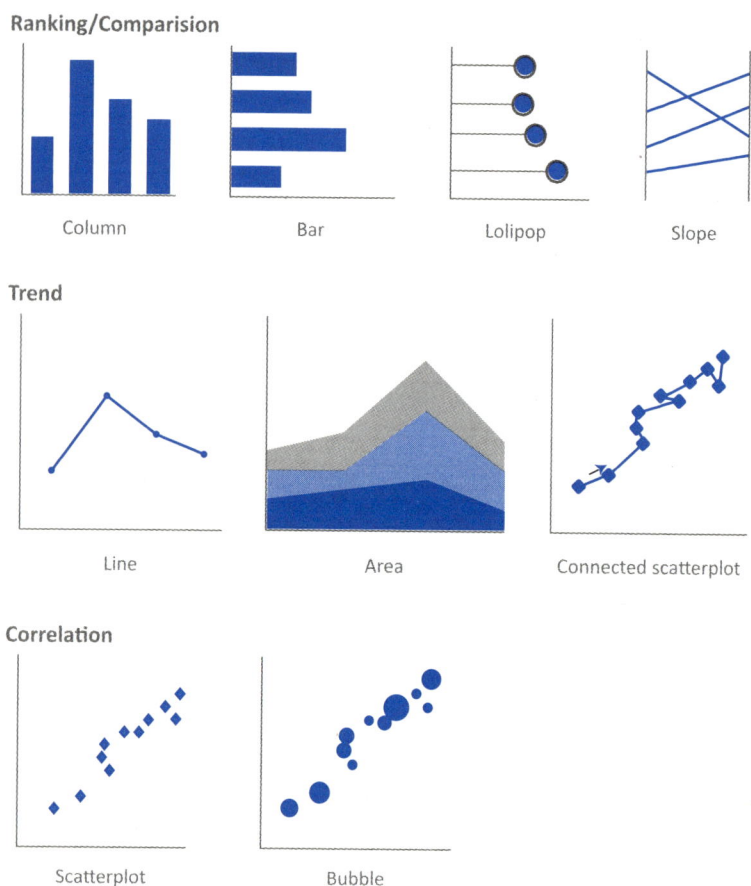

Figure 1.5: Categorization of data charts.

top to down. The other scenario where bars work better than columns is if the individual category names contain many characters. A column chart in this case may land up looking quite messy, while a bar chart provides ample space on the y-axis for lengthy category names.

Another construct which can be used for comparing values is a **lollipop chart**. Lollipop charts can be attributed to Andy Cotgreave, a visualization designer working for Tableau

Distribution

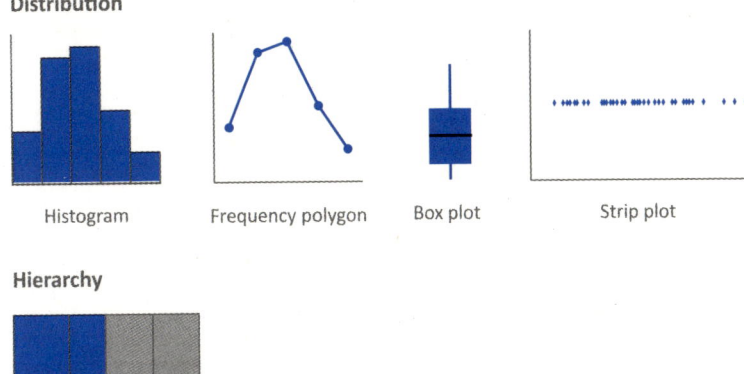

Histogram   Frequency polygon   Box plot   Strip plot

Hierarchy

Tree map

Composition

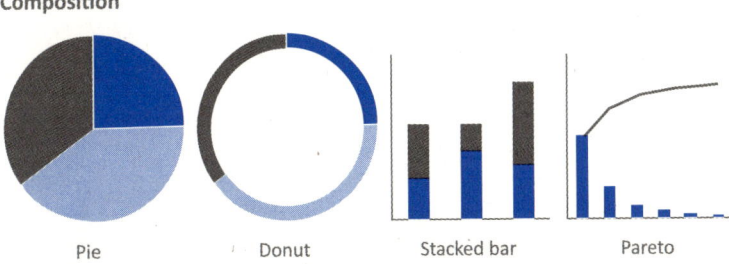

Pie   Donut   Stacked bar   Pareto

Figure 1.6: Categorization of data charts (part 2).

(a popular business intelligence [BI] tool). Why would one use lollipop charts instead of bars or columns? It could be used for a variety of reasons, including reducing moiré vibrations[2] if there are many large bars next to each other, adding variety to your dashboard or reducing data ink to make your graph more pleasing to the eye. A full discussion on the merits/demerits of

---

[2] Moiré vibrations is a term used to describe the dizzying or uneasy feeling produced by lines or patterns of contrasting colours.

lollipop charts and how to create them is covered well in an article[3] by Ben Neville, which is available online.

Along with bars, columns and lollipops for comparison come **slope graphs**. These are a special case of line graphs which show only two data points for each category, typically a 'before' and 'after'. Introduced by Edward Tufte in what is generally considered his best book *The Visual Display of Quantitative Information*, the slope graph allows the comparison of values at two different time points. For example, the per capita GDP of different countries in 2009 versus 2019 could be represented with a slope graph, with a line for each country. The slope of each line would indicate the rate of change of the variable (the per capita GDP in this case), and the start and end points of each line would indicate the position and ranking of countries at the two timestamps.

Trends

Line Chart | Area Chart | Connected Scatterplot

A trend or time series relationship captures the behaviour of a variable over time. Does the variable increase or decrease over time? Are there any cyclic patterns? Are there any sudden spikes or drops, for example, the performance of a company's stock prices over the last three months?

When trends need to be visualized, 'line charts' work best. As we all know, in a typical line chart, the x-axis is used for the time scale and the y-axis is used for the quantitative variable (or variables) being depicted. There is no alternative for this configuration (with the lines flowing from top to down, for example) as we have in the case of bars and columns. Have you ever wondered why? This is because we seem most comfortable

---

[3] https://www.tableau.com/about/blog/2017/1/viz-whiz-when-use-lollipop-chart-and-how-build-one-64267

with visualizing the passage of time from left to right, rather than top to bottom, bottom to top or right to left. This may at least partially be attributed to our conditioning of reading text from left to right. There are rare cases of line or bar charts with time flowing from right to left, but these tend to confuse or mislead the poor reader who happily assumes that time is flowing in the direction it usually does! Keep this in mind the next time you create a chart which has time as one of the axes (more about this in Chapter 4).

Area charts are also often used to depict trends, especially when the behaviour of multiple variables needs to be compared over time along with depicting the total of these variables over time. However, area charts are ripe for hiding the actual trend of a variable, especially if it is stacked on top of other variables. We shall study this aspect of area charts further in Chapter 4.

Finally, in this category, we have connected scatterplots. These build on the traditional scatterplot by joining all the dots chronologically. A scatterplot as discussed in the next subsection consists of two quantitative variables on each axis and a dot to represent each reading. In a connected scatterplot, a temporal dimension is added by chronologically joining all the dots, so that viewers can discern how the relationship between the two variables has changed over time. Since the line could be zigzagging to the right, left, up or down, it is important to include arrows in the connected scatterplot to clearly depict the direction of the flow of time. Omitting this could land up confusing the reader as discovered by Haroz et al. in their study[4] on the effectiveness of connected scatterplots.

---

[4] Steve Haroz, Robert Kosara, and Steven L. Franconeri, 'The Connected Scatterplot for Presenting Paired Time Series', *Transactions on Visualization and Computer Graphics* 22, no. 9 (2016): 2174–2186.

Correlation

Scatterplot | Bubble Chart

The relationship between quantitative variables or, in other words, the correlation of two or more quantitative variables is a data relationship which is commonly studied, for example, the monthly average temperature and total monthly ice cream sales. The most relevant questions here are whether and how one variable is related to another. As one variable increases, does the other increase or decrease or remain the same? A positive correlation means that both increase together (though not necessarily at the same rate). A negative correlation implies that one variable increases as the other decreases.

The most institutive display for studying a correlation between two variables is the scatterplot which uses both the x and y-axes to display one quantitative variable each. In our example of temperature and ice cream sales, each dot on the plot would depict data for one month. If indeed we noticed that there was a positive correlation between the two variables (i.e., more ice cream was sold in months with higher temperatures and lower temperatures coincided with lesser ice creams being bought), we could venture that there was a positive correlation between ice cream sales and temperature. However, our posturing would have to end there, and we would be fraught to conclude that hotter weather was making people buy more ice creams. What if actually there were some holiday discounts which were offered in summer which enticed people to buy more ice creams? A correlation as seen here does not imply a causation. Hence, a scatterplot remains a tool which helps us notice correlations between two variables, and we should refrain from jumping to casual 'causal' conclusions without doing adequate work to justify causality in either direction.

A bubble chart is a special kind of scatterplot, where instead of two, three quantitative variables can be depicted on the same chart. The first two variables are depicted on the x and y-axes (as in a normal scatterplot), and the third variable is captured by the size of the dot, which now becomes a 'bubble'. In our earlier example of temperature and ice cream sales, we could introduce a third variable called 'discounts', which would be depicted by the size of the bubble for each month. It should be noted that size variations are more difficult to judge accurately than one-dimensional (1D) positions on an axis. Hence, changes in the third variable here will inherently be more difficult to access than the first two. In spite of this limitation, bubble charts are quite widely used to capture three quantitative variables in the same view.

Distribution

Histogram | Frequency Polygon | Box Plot | Strip Plot

A distribution relationship is when we want to know how a quantitative value is spread across a whole range, for example, distribution of end-term marks for a particular course. The range here could be from zero to the total possible marks. Often, there might be too many data points to plot individually, and it might make more sense to chop up the full range of the data into bins and count the number of occurrences in each bin. This leads to what is commonly called a frequency distribution. In our example, suppose the course has 500 students, and the marks range from 0 to 100; we might decide to create 5 bins with a marks range of 20 each. The count of students in the first bin would denote how many earned less than 20 marks each, the next bin would denote how many earned between 20 and 40 marks and so on, giving us a frequency distribution of how the entire class has performed in the end term.

Histograms are a popular and effective visual representation of a frequency distribution, provided the bin size is chosen carefully. A histogram is structurally very similar to a column chart, except that each column represents a bin and the design dictates that there is no gap left between the columns. Take a moment to think about why there are no gaps between the columns in a histogram. This is intentionally done to denote the continuity in the ranges of the different bins. Remember that one continuous range was purposefully chopped up into distinct bins to understand the distribution of the data. The design of the histogram helps us remember this subtle but important fact.

Frequency polygons are to line charts what histograms are to column charts. In other words, we might choose to depict the shape of the frequency distribution by using a line instead of columns. Frequency polygons work better than histograms if we need to plot multiple frequency distributions in the same view to possibility compare them. For example, you as a college professor have a vague notion that students have been performing better in your course over the years. You could use a frequency polygon (with one line for each year of data you have) to see if this is indeed true. However, if you had data for more than four or five years, the overplotting might make your frequency polygon look more like a noodle bowl! In situations like these, with many individual distributions to compare, box plots might come to your rescue.

The box plot was invented in the 1960s by John Tukey, a Princeton mathematician who is also known as the father of exploratory data analysis. Tukey specifically created the box plot (also called the box-and-whisker plot) as a means to concisely represent a distribution in quartiles. Given that it is a dense representation of a distribution, the box plot lends itself beautifully for comparisons across multiple distributions, where each box plot represents one distribution. For a short but effective introduction to box plots, refer to the article by Nathan Yau on flowing data.[5]

A word of caution: Box plots are not as intuitive as other basic charts we have discussed so far. If your audience is not familiar with the box plot, remember to factor in a couple of extra minutes in your presentation to first explain the construct before you use it for presenting your data. Similarly, if you choose to use this construct in a report, make sure to point the reader to some material on box plots for ready reference (or even better, include it in your report as additional reading).

Another construct quite useful for depicting a distribution is a strip plot (also known as the dot plot). Often referred to as the 'one-dimensional scatterplot', a strip plot has only one axis and marks each value in the dataset as a dot. Unlike the histogram or frequency polygon, strip plots do not capture the shape of a distribution but are nonetheless helpful for precise depictions of small datasets. For larger datasets, overplotting in strip plots can be avoided in two ways (see Figure 1.7):

1. Make the points transparent so that darker regions denote denser data distribution.

2. Add jitter to the points (shift points vertically if the x-axis contains the quantitative value or vice versa).

Hierarchy

Tree Map

Created in the 1990s by Ben Shneiderman, a computer scientist at the University of Maryland, the original purpose of the tree map was to visualize the contents of a hard drive. A typical hard

---

[5] https://flowingdata.com/2008/02/15/how-to-read-and-use-a-box-and-whisker-plot/

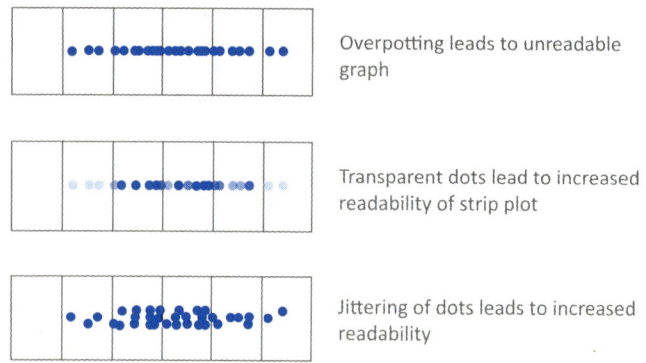

Overpotting leads to unreadable graph

Transparent dots lead to increased readability of strip plot

Jittering of dots leads to increased readability

Figure 1.7: Countering overplotting in strip plots.

drive could have tens of thousands of files organized in hierarchical folders 5–15 levels deep. Depicting this data while maintaining the hierarchical relationship of different categories is tricky. A tree map would be organized in rectangles, with bigger rectangles comprised of smaller rectangles and so forth. In the example of the hard drive, each leaf rectangle would represent a file, with the area of the rectangle directly proportional to the file size. The colour of each leaf rectangle could denote the type or category of the file. Each bigger (composite) rectangle would represent a folder, with higher-level folders composed of lower-level folders as rectangles and so forth.

Tree maps are now used to visualize a wide range of hierarchical data from sales of food items in different categories to stock performances of different companies in different sectors. For further details on how to use tree maps, refer to Shneiderman's article[6] on perceptualedge.com.

---

[6] Ben Shneiderman, 'Discovering Business Intelligence Using Treemap Visualizations' (2006), http://www.perceptualedge.com/articles/b-eye/treemaps.pdf

Composition
Pie Chart | Donut Chart | Stacked Bars | Pareto Chart

Composition or part-to-whole relationship is used when we want to show how individual categories are contributing to the total, for example, contribution of revenue from different products in our portfolio to the total revenue. Composition relationships are usually better understood if they are presented as percentages.

The most popular representation of the composition relationship is the pie chart. Pie charts have been in wide use in both popular press and business scenarios. However, pie charts have had a fair amount of criticism come their way with visualization experts across the board from Edward Tufte to Stephen Few calling out their ineffectiveness. Box 1.1 contains some quotes on pie charts from well-known visualization experts.

Why are pie charts frowned upon by so many visualization experts? Well, study after study has shown that bar charts and sometimes even simple tables work better than pie charts for

'Pie charts [along with other forms of area charts] do not provide efficient detection of geometric objects that convey information about differences of values.' William Cleveland

'Save the pies for dessert.' Stephen Few

'The only worse design than a pie chart is several of them.' Edward Tufte

'We make angle judgments when we read a pie chart, but we don't judge angles very well.' Naomi Robbins

'Pie charts are evil.' Cole Nussbaumer Knaflic

Box 1.1: What data visualization experts have said about pie charts.

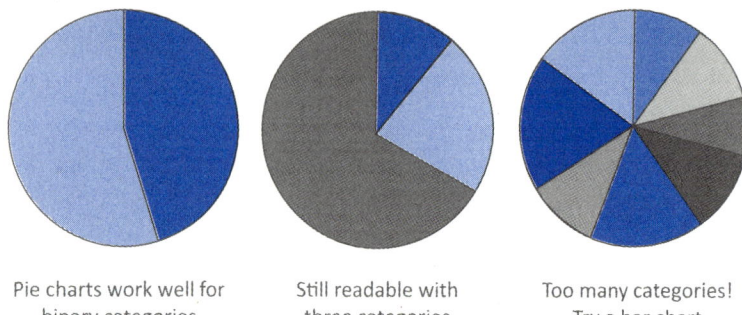

Pie charts work well for binary categories

Still readable with three categories

Too many categories! Try a bar chart

Figure 1.8: Pie chart recommendations.

accurately interpreting values. Pie charts require us to be able to accurately decode values of angles and/or areas. It just so happens that humans are better at decoding values if they are presented as length (bar graph) or position (dot plot) instead.

Does it mean that there should be no place for the pie chart in your graphical toolkit? Not necessarily. Following are two specific cases where I think pie charts still hold a place of honour.

1. If the number of categories you need to represent is small, especially if there are only two categories (refer to Figure 1.8).

    As Nathen Yu puts it, pie charts with up to four slices are 'still bearable',[7] but beyond that, as you increase the number of slices, your chart gets more and more unreadable. I would suggest that you limit the number of slices in your pie chart to at most three.

2. If you want to show that a group of categories adds up to less than the sum of the remaining categories.

We will revisit pie charts along with finer design aspects for creative effective pies in Chapter 5.

---

[7] https://flowingdata.com/2015/08/11/real-chart-rules-to-follow/

Donut charts are similar to pie charts, as both use a polar coordinate system. At least one study[8] has found donut charts to be as effective or ineffective as pie charts, so use them with the same amount of caution you would apply to a pie chart.

Stacked bar graphs are often used as an alternative to pie charts when a part-to-whole or composition relationship needs to be displayed. A stacked bar works better than a pie chart in estimating values because the data is now depicted in one dimension (as length of the bar) instead of two dimensions (as area of a slice). However, the individual categories are still difficult to compare because a stacked bar lacks a common baseline for all the categories. Which construct represents values in one dimension and uses a common baseline? You guessed it right—the plain old bar or column graph.

Stacked bars, however, do work well for a particular use case—when you need to show changes in the percentages of categories (perhaps over time). Even in this case, the construct is most useful if the number of categories is limited to two or at most three. Figure 1.9 shows how 100 per cent stacked bars can be used to show changes in composition over time. With two common baselines to work with, we can clearly see the change in each category over time.

The Pareto graph is a simple column graph with an added feature—a cumulative line spanning horizontally, representing percentage contribution of all categories till that point. Pareto graphs get their name from Vilfredo Pareto, an Italian economist who coined the Pareto principle, also known as the 80/20 rule. Legend has it that he noticed that 80 per cent of the land in Italy was owned by 20 per cent of the population. He then went on

---

[8] Drew Skau and Robert Kosara, 'Arcs, Angles, or Areas: Individual Data Encodings in Pie and Donut Charts', *Computer Graphics Forum* 35, no. 3 (2016): 121–130.

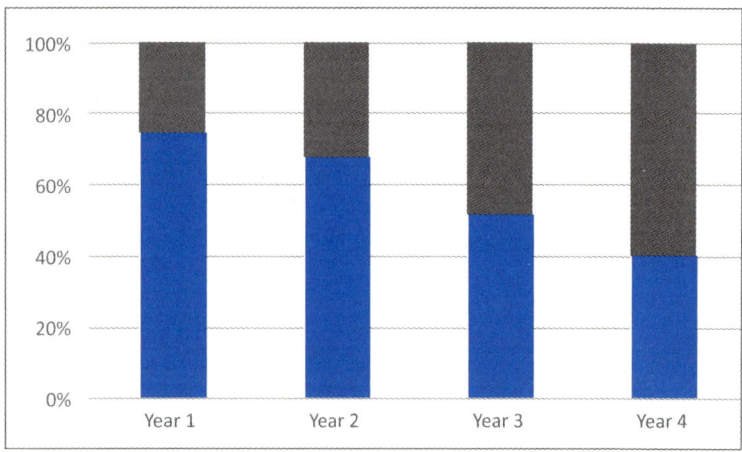

Figure 1.9: A 100 per cent stacked bar graph contains two common baselines (top and bottom), allowing us to accurately see how both categories have changed over time.

to realize that 80 per cent of the peas in his garden were harvested from 20 per cent of the plants. The Pareto principle generalizes this phenomenon by stating that for many events, 80 per cent of the results are caused by 20 per cent of the effects.

Pareto graphs are a great choice when you want to highlight the 80/20 phenomenon (or a similarly skewed ratio) in your data. For example, 85 per cent of our sales come from 25 per cent of our clients.

With this, we conclude our quick tour of graphical constructs which are commonly used.

NAVIGATING THE BOOK

This book is organized under different major themes of how graphs can go wrong in what information they represent and convey. Each chapter tackles a particular theme with myriad examples of misleading graphs and walks the reader through alternate, more truthful ways to depict the data and convey the intended message. While each chapter is fairly independent, there are certain places where the book refers back to something we dealt with in an earlier chapter. Hence, to the extent possible, it is recommended that you read the book in sequence, though I can understand the temptation to jump to examples that catch your eye first (perhaps because they strike closer to home?).

Chapter 2: Representing Quantities Truthfully

The first principle in creating a truthful graph states that the amount of ink used to represent a quantity should be in proportion to the quantity it represents. Bar graph scales which do not start at zero, using the diameter of a circle to represent a quantity and disproportional icons, all flout this simple rule. These and other common mistakes in plotting data are discussed in Chapter 2, along with alternate encodings which are more accurate and hence more effective.

Chapter 3: The Deception of the Third Dimension

Three-dimensional (3D) charts are a bad idea. Using the third dimension for a chart can not only make the chart difficult to read but can actually force us to interpret the data incorrectly. This chapter introduces various problems 3D constructs can create. We simultaneously explore alternate graph constructs for each of these problematic scenarios and effective solutions to capture multiple dimensions on two-dimensional (2D) media.

Chapter 4: Spurious Trends and How to Spot Them

Why do visualization experts frown upon dual axis graphs? From manipulating the scale of the axis to changing the aspect ratio of a plot or not providing enough context, this chapter deals with various tricks commonly used to misrepresent trends. Drawing from real-world examples, we look at how even some very reputed sources have succumbed to these ploys. Most importantly, the chapter emphasizes design guidelines which will help you visualize trends correctly.

Chapter 5: Design Choices for Accurate Data Interpretation

The devil as we all know is in the details. How does the choice of a particular colour palette or the position of a label have the potential to create a misleading graph? This chapter shows you the importance of paying attention to the finer design aspects of your visualization. Graphical experts leverage many of these basic design principles to create effective visuals and now you too can do the same.

Chapter 6: Tell Your Data Story

How do you make sure that the graphs you present make sense to your audience and create the desired impact? This chapter covers principles of visual perception which can be leveraged to get your audience to notice a particular message in the data and hence help you create an effective data narrative. It also alerts you to certain practices which could distract the audience from the actual data. This chapter also contains exercises to practise your newly acquired skills on plotting data accurately.

CHAPTER 2

REPRESENTING QUANTITIES TRUTHFULLY

Graphic excellence begins with telling the truth about the data.

Edward Tufte,
The Visual Display of Quantitative Information

A simple yet fundamental rule of thumb (introduced by Tufte in his seminal book *The Visual Display of Quantitative Information*) to help maintain the integrity of your graph is to make sure that the amount of ink used to represent different quantities is directly proportional to the quantities themselves. Violating this rule is the easiest path to a deceptive graph. The most common example of a graph that flouts this rule is a bar chart that has a non-zero baseline.

BAR CHARTS AND HOW NOT TO TRIP ON THEM

Figure 2.1a uses a bar chart to depict the revenues for three companies. Notice how we can intuitively deduce that Company 2 has double the revenue of Company 1. Now notice what happens if we make a small change to the bar graph as in Figure 2.1b. The y-axis scale of the graph now starts at 1 instead of 0. A quick glance at only the second graph could lead us to wrongly conclude that Company 2's revenue is around thrice that of Company 1 because the bar for Company 2 is three times the length of Company 1's. To avoid such misleading illustrations, the length of the bars should always be in direct proportion to the quantities they represent. Bar graphs where the axis' scale does not start at zero flout this basic rule.

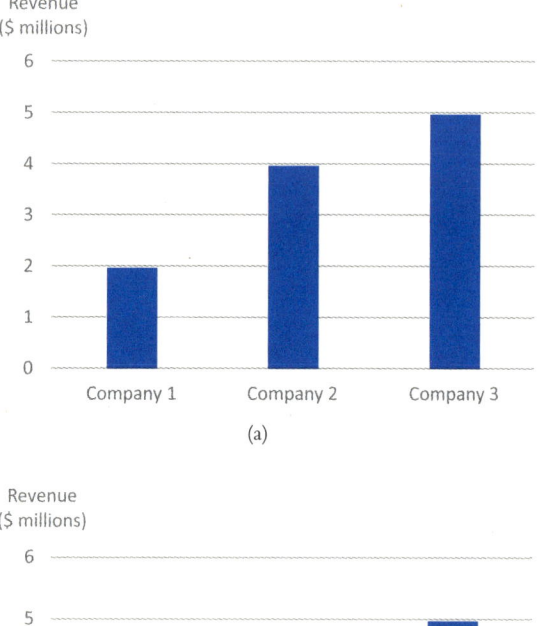

(a)

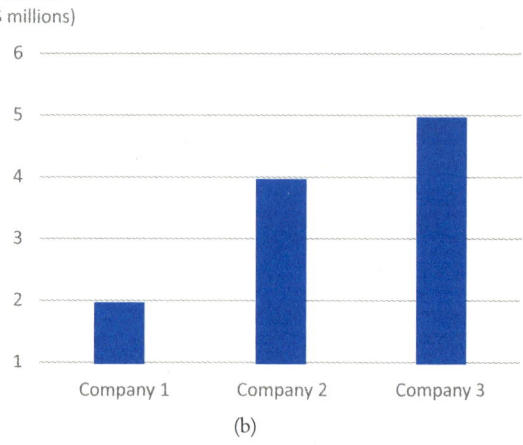

(b)

Figure 2.1: Bar graphs where (a) the y-axis scale starts at zero and (b) the misleading version where it does not.

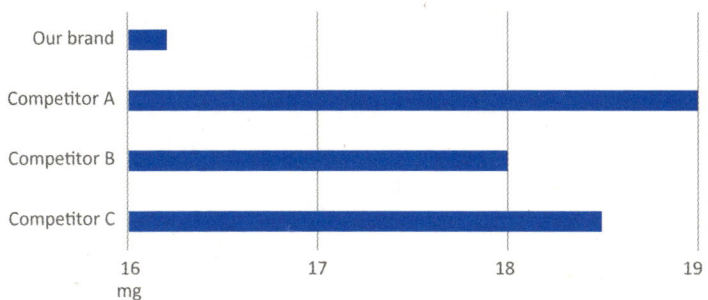

Figure 2.2: Original misleading version with truncated scale.

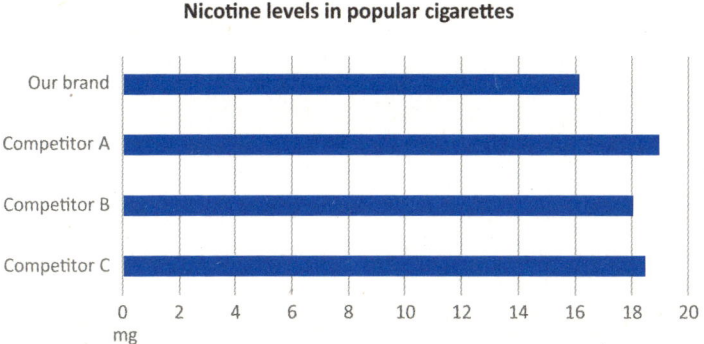

Figure 2.3: Corrected version.

There are examples abound where such bar graphs have been intentionally used to convey a message in favour of a particular narrative. Let us revisit the example we looked at in Chapter 1 which depicted nicotine levels in different cigarette brands. A similar graph was indeed used by a tobacco company in the advertisements of their product. Looking at both versions together (Figure 2.2 is the truncated axis version versus the non-truncated axis in Figure 2.3), one can clearly see the subterfuge in the original version in which 'our brand' comes out looking far less lethal than the competition. In truth, the nicotine levels in 'our brand' of cigarettes is only marginally lesser than the competition.

The by now notorious bar graph used by Fox News[1] to depict the sign-up numbers for ObamaCare (the healthcare programme sponsored by Barak Obama, the then president of the USA) used a similar tactic. Six million users had signed up for the programme, which had an ultimate target of seven million, denoting a successful launch. However, the bar graph depicting these two values used a truncated axis which showed a ratio of 1:3 (instead of 6:7) between sign-ups and target, indicating a failure of the programme.

The counterargument or justification often given for the truncated axis bar graph is often as follows—starting the quantitative axis scale at zero makes all the bars look similar in length, thus making it difficult to discern small differences in values.

While the argument does hold some merit, if the goal is to discern small differences between values, perhaps a bar chart is not the right choice. Consider a simple table or a dot plot instead. Either would be adequate to bring out small differences in large values without creating a misleading visual. Note that for a dot plot, it is acceptable for the quantitative scale to start at any value and not necessarily at zero. This is justifiable because the dot plot is read by judging the value each dot represents rather than its distance from the baseline. In contrast, in a bar or column graph, the actual length of a bar is meant to depict a quantity.

If a bar chart is imperative for your visualization, one way to bring out small differences in values is to use labels on the bars,

---

[1] https://mediamatters.org/fox-news/dishonest-fox-charts-obamacare-enrollment-edition

creating what is often called in the data visualization community a 'grable', that is, a **gr**aph + a t**able**.

Figure 2.4 depicts both these alternatives for accentuating small differences in your data.

Another approach which invariably leads to a misleading bar graph is when the quantitative scale is broken (see Figure 2.5). The temptation to break the quantitative scale often arises when

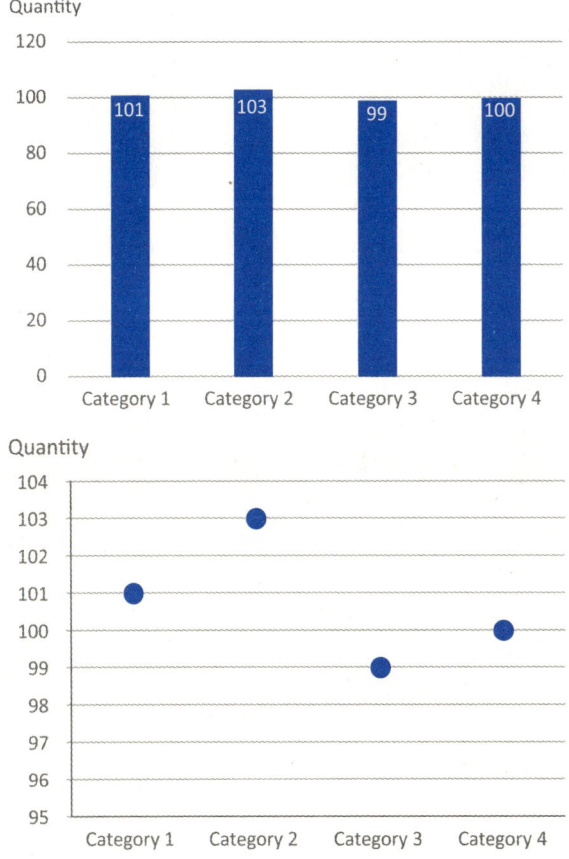

Figure 2.4: A labelled bar graph and a dot plot as options to accentuate small differences in quantities.

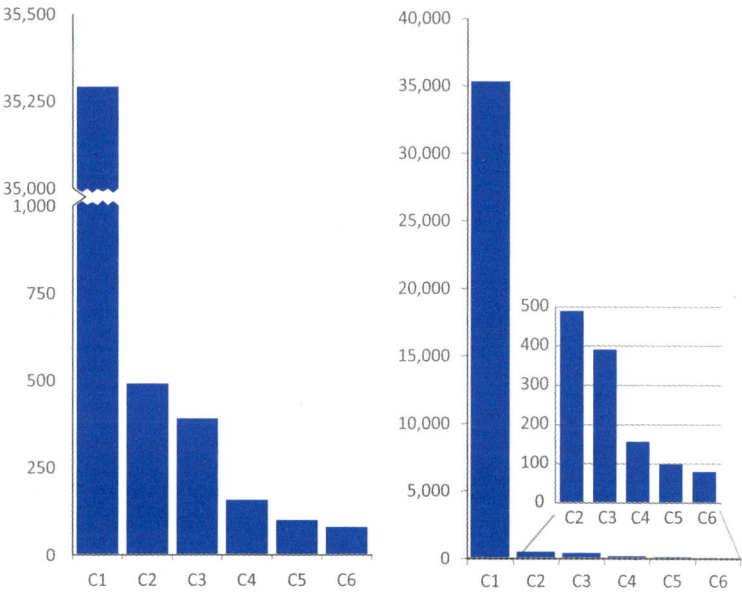

Figure 2.5: A break in the axis creates a misleading chart (left) while the combination chart on the right remains accurate to the data while enabling comparisons between all categories.

there are large differences in the quantities which need to be depicted in the same graph. With a continuous axis scale, the smaller categories all land up looking the same as can be seen on the graph on the right in Figure 2.5, making it impossible to compare them to each other. However, a break in the axis scale leads to a graph where the bar lengths cannot be compared to each other for any meaningful insights.

The graph on the right in Figure 2.5 depicts a practical workaround to the situation—do not tamper with the scale but use a blowout of the smaller categories with a different scale to enable their mutual comparison as well.

The next question that arises when we talk about variations in the axis is regarding the log scale. When is it useful to use a log scale for a quantitative axis? Does using a log scale create a misleading graph?

Log scales are generally helpful when there are large variations in the data and they need to all be fitted into the same graph. See the example in Figure 2.6 and judge for yourself if the log scale creates a misleading graph.

Recall the rule defined in the beginning of this chapter—the amount of ink used to depict the object should be proportional to the quantity it represents. This rule is definitely violated when log scales are used.

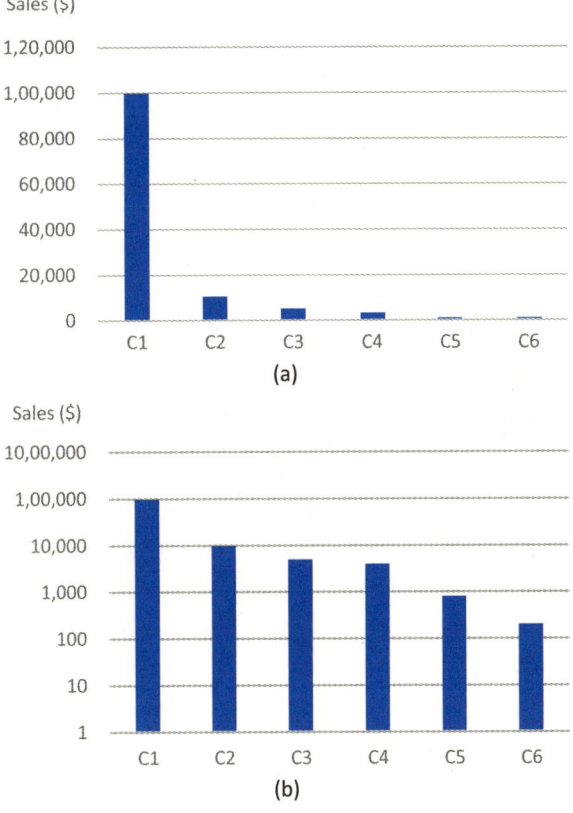

Figure 2.6: Same quantities represented on a linear scale (a) and log scale (b).

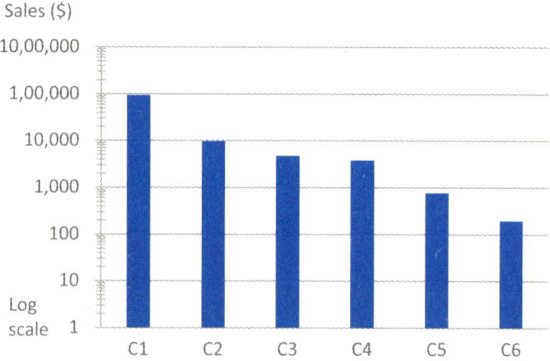

Category wise sales for last quarter (log scale)

Figure 2.7: Elements like minor tick marks, an axis label and a clear title alert the user that the graph has a log scale.

There are two possible scenarios when a log scale could trip an unsuspecting user—the user either assumes that a linear scale is in use or is aware that the scale is logarithmic but does not know how to interpret the graph. In either case, there is a danger of the data getting misinterpreted. In the example shown in Figure 2.6, the fact that C1 is at least one order of magnitude larger than the largest of the other categories is visually obscured when a log scale is used.

If, however, you have a compelling reason[2] to use a log scale in your graph, make sure to alert the user of the same. This can be done by including a few additional elements in your graph such as minor tick marks (which look very different for a log scale as compared to a linear scale), a clear axis label and/or an alert in the title of the graph itself. The example in Figure 2.7 uses all these elements to make sure the reader is cautioned in multiple ways that a log scale is in use.

[2] Harvey Motulsky, 'The Use and Abuse of Logarithmic Axes' (2009), https://cdn.graphpad.com/faq/1487/file/1487logaxes.pdf

SCALES FOR OTHER CHART TYPES

While the zero baseline rule is non-negotiable for a bar chart (and by corollary the lollipop chart), it does not apply to a line chart or a scatterplot.

In a line chart, the primary message is conveyed by the relative horizontal movement (or trend) of the line. The height of the points on the line are of secondary importance. There are instances where starting a line graph at zero might impend the reading of the graph, and in such instances, it is perfectly acceptable for the y-axis scale to start at any other convenient value. Figure 2.8 shows one such example, where starting the base at zero causes overplotting of data and does not allow the user to clearly discern the behaviour of different categories. Fixing the scale of the y-axis according to the range of the data introduces more vertical space, allowing the lines to be visually distinct.

Sometimes starting the y-axis scale at zero for a line graph can actually perpetuate misperceptions. With a y-axis starting at zero, small differences in values can get hidden, which could be problematic for metrics like global temperature where even small differences could signify monumental changes.

A graph on global temperature over time, which was tweeted by the *National Review,* used a scale, surprisingly starting not even at zero but at −10 degrees (F) to plot Earth's temperature across decades.[3] While the temperatures in the graph are well within the range of 56–58.5 degrees (F), the axis in the graph ranges from −10 to 110, effectively showing a flat line. Environmental experts unilaterally agree that a change of more than 1.5 degree is dangerous for the planet, which means that an axis ranging

---

[3] https://www.researchgate.net/figure/A-tweet-by-National-Review-on-December-14-2015-showing-the-change-in-global_fig1_321757804

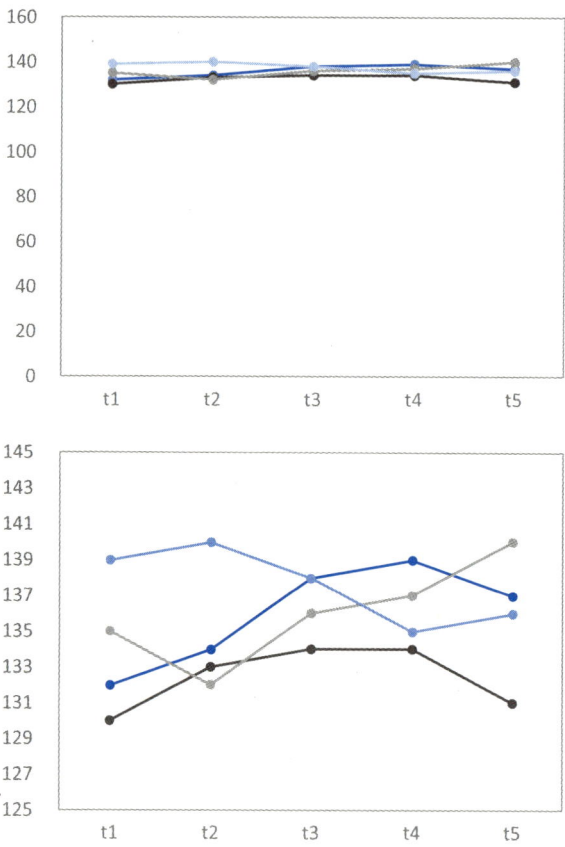

Figure 2.8: A line chart with a zero baseline becomes more readable after the scale of the y-axis is adjusted according to the range of the data.

between 56 and 59 would have been the truthful way to show the sharp spike in recent global temperatures. The tweet was deleted after it caused an uproar, but the agenda of the *National Review* graph is clear for anyone to see—to label climate change a hoax. Ultimately, the scale of a line graph should be chosen considering the domain and context of the data.

There is one exception, however, to the guideline just discussed. If the area below the line is shaded, then the line chart is no longer a line chart but becomes an area chart. In this case, the height of the line has a significance, and hence the y-axis scale 'should' begin at zero—if it doesn't, then a misleading graph is created.

The example in Figure 2.9 depicting monthly precipitation in Singapore illustrates this case. In the first graph, the reader would not be at fault if they concluded that June, July and August are extremely dry months in Singapore and November/December put together get the bulk of the precipitation. However, this graph is misleading because of the truncated y-axis scale. Once the baseline is corrected to zero (second graph in Figure 2.9), it is quite clear that Singapore experiences significant precipitation throughout the year.

Similar to a line chart, for a scatterplot, judicious choosing of the starting point for both quantitative axes scales is a crucial design step for creating a meaningful visual. As can be seen in Figure 2.10, starting the scales at zero in this scatterplot is quite ineffective. Not only is there a massive waste of space, the data points are clustered together and are difficult to interpret. Adjusting the scales to start at where the data starts allows us to clearly see the positive correlation between temperature and ice cream sales.

If the gentle reader is a bit confused after these meanderings, here is a quick recap on the design of axes: bar, column, lollipop and area charts should always have a quantitative scale which starts at zero, while the quantitative scales for line charts, dot plots and scatterplots can start at any appropriate value (including zero) which allows the data to be plotted clearly.

ICONS IN BAR CHARTS

Our conversation on bar charts would be incomplete without discussing the use of icons in graphs. Icons or pictographs are

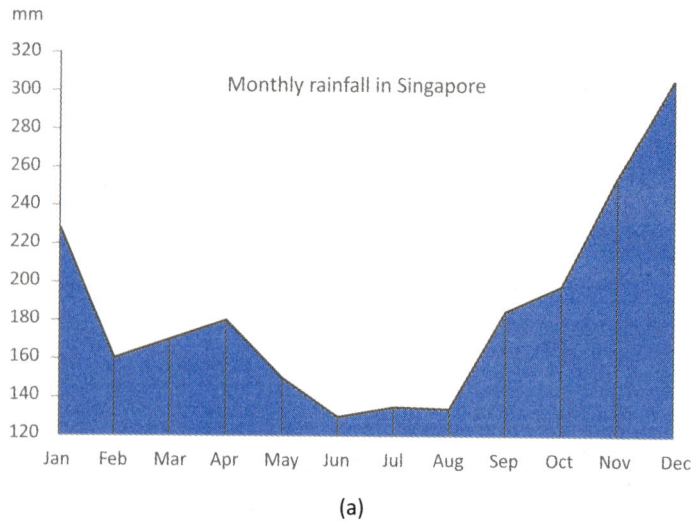

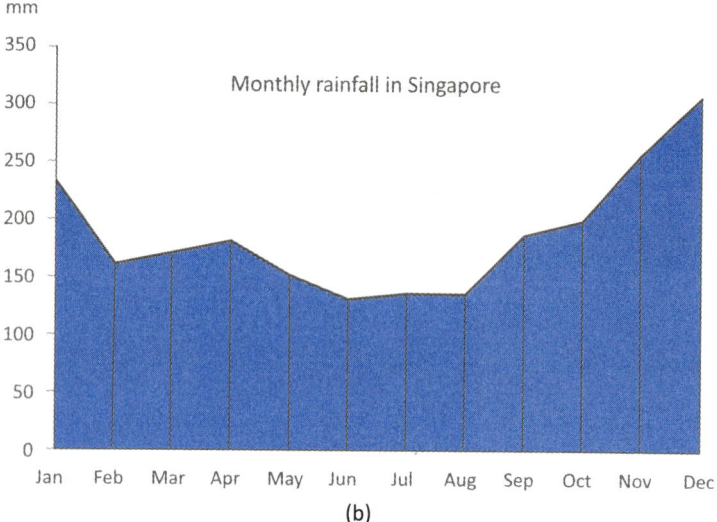

Figure 2.9: An area chart with the y-axis scale not beginning at zero (a) is misleading till the scale is corrected to start at zero (b).

commonly used in bar charts to represent the frequency of different entities. Improper scaling of icons can create a distorted graph as in the example on snack preferences in a particular 3rd grade classroom, depicted in Figure 2.11.

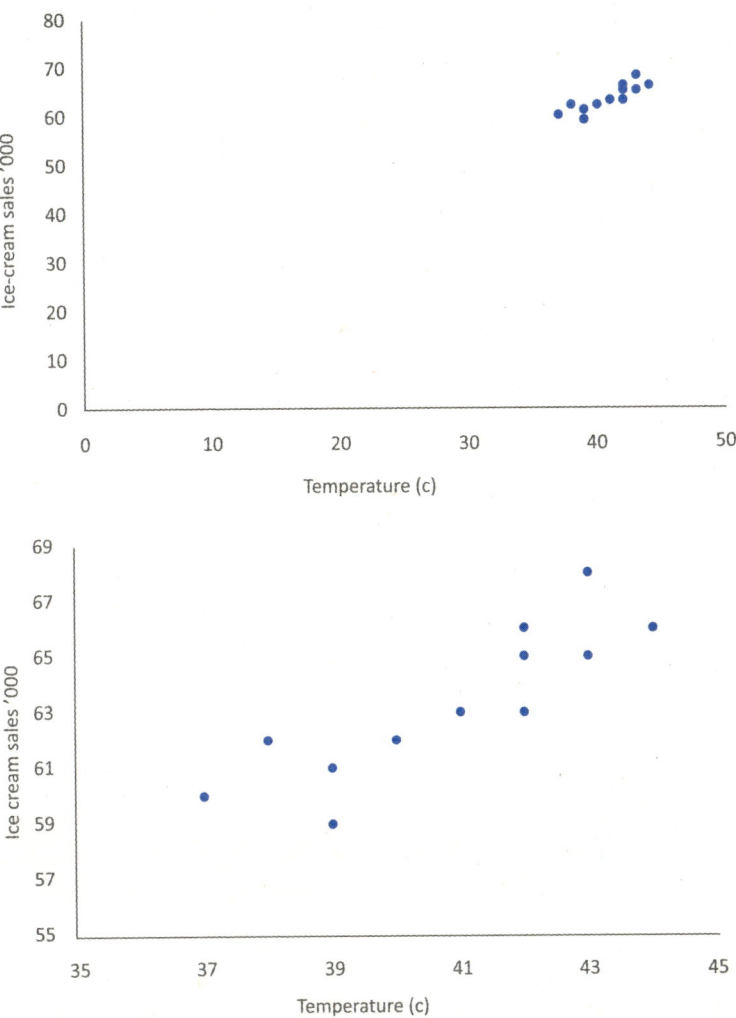

Figure 2.10: Space wastage and overplotting in a scatterplot can often be rectified by choosing appropriate axes ranges.

Why is the chart on the left misleading? Notice how popcorn seems the most popular snack in the first chart (judging by the length of the popcorn 'bar') but is actually less popular than ice cream cones. While there are six ice cream icons, there are only

Preferred snack among 3rd graders   **Preferred snack among 3rd graders**

Figure 2.11: Icons with different widths create a misleading graph (left) which is corrected (right) by making all icons of uniform width.

five popcorn icons (with the assumption that each represents one vote). Similarly, tacos seem more popular than cheese-cake, but there are actually three of each. The chart on the right corrects the visual by making sure each icon is of the same width and all icons are vertically aligned. In the corrected version, it is easy to accurately identify ice cream as the most popular snack.

Another common slip-up while using icons is improper 2D scaling as seen in the Figure 2.12. Cheese-flavoured popcorn has double the sales of plain popcorn, but scaling up the icon to twice the height increases the width as well. We land up with a cheese popcorn tub which is visually four times larger than the plain popcorn tub!

In this case, the correct way to represent twice the amount of popcorn is simple— replicate the icon without scaling it (graph on the right). Suppose cheese popcorn sales were one and a half times that of plain popcorn, half a tub of the same size could be depicted to capture the half. However, the alert reader might have realized that going down this path could quickly get quite

Figure 2.12: Improper scaling of the icon (left) leads to a misleading graph where cheese popcorn sales seem to be four times that of plain popcorn instead of only twice. The graph on the right corrects the illusion.

messy. What if the ratio is 1:1.8? How would you visually distinguish that from say a ratio of 1:1.9?

Given the above scenario, it is safe to assert that icons in graphs are useful in a limited set of scenarios and should be used only if there is a compelling need for the same.

CIRCLES FOR VALUES: MYSTERY SOLVED!

Scaling issues are more common than we would imagine. The chart in Figure 2.13 (recreated for clarity) appeared in the *New York Times* article titled 'The Wild West of Finance'.[4] The article talked about how the biggest US banks had grown larger and the top three banks had a huge chunk of the market share—close to 44 per cent. Look at the circles in Figure 2.13 used to represent these numbers. Does the blue circle look like 44 per cent of the largest (dark grey) circle? The blue circle is actually closer to 20 per cent rather than 44 per cent of the largest circle, which does a huge disservice to the message the article wanted to convey!

---

[4] https://www.nytimes.com/2011/12/11/magazine/adam-davidson-wild-west-of-finance.html

All banks

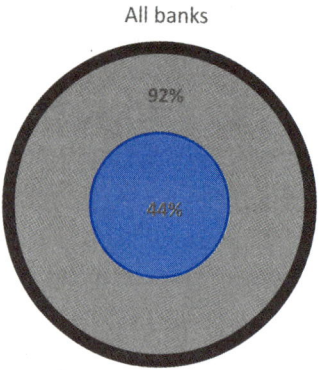

92%

44%

Top 3 banks
Top 20 banks (including top 3)

Figure 2.13: Chart from *The New York Times* which uses the diameter to encode values instead of the area. The blue circle actually occupies around 20 per cent of the area of the biggest circle and not 44 per cent as was intended.

So what exactly happened here? It turns out that the creators of the chart used the diameter of the circles to represent the ratios of 44, 92 and 100, which dramatically skews up the ratios of the circles because the diameter is squared to get the area.

Figure 2.14 (right) contains a corrected version of the chart, and you can now see how much larger the blue circle needed to be, to accurately represent 44 per cent of the big circle.

Figure 2.15 compares the relative sizes of circles for area versus diameter encoding. As is evident, the squaring of the diameter ensures that the diameter encoded circles highly distort the actual ratios they are supposed to represent. For example, the largest circle in the bottom row is supposed to be only 6 times larger than the smallest circle in the same row but is actually 36 times larger. Note that its diameter is six times larger, but the viewer (justifiably) tends to compare the areas.

Using the diameter to encode values happens to be such a common error in graphing data (intentional or otherwise) that

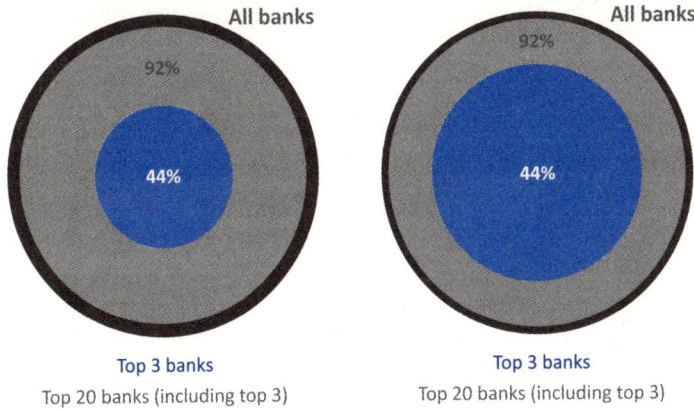

Figure 2.14: Inaccurate (left) and accurate (right) representations of the market share of US banks.

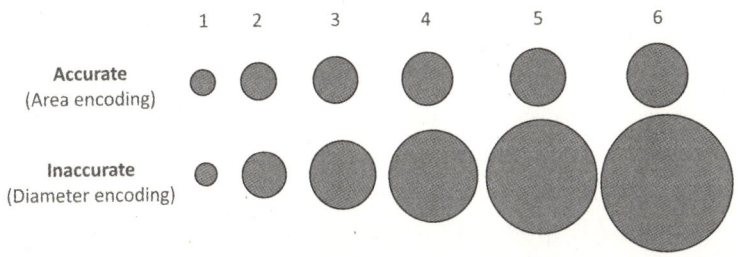

Figure 2.15: Comparison of area encoded versus diameter encoded circles.

many data visualization experts including Stephen Few, Alberto Cairo and Naomi Robbins[5] have brought attention to it.

Stephen Few, in his article titled 'Our Irresistible Fascination with All Things Circular',[6] discusses a data visualization which

[5] https://www.forbes.com/sites/naomirobbins/2012/02/28/misleading-graphs-displaying-a-change-in-one-variable-using-area-or-volume/#6886d1d81781

[6] Stephen Few, 'Our Irresistible Fascination with All Things Circular', *Visual Business Intelligence Newsletter* (2010), http://www.perceptualedge.com/

was put out by JPMorgan which uses circles to encode the market value of their own company and some others in the same sector. The visual lands up highly exaggerating the loss of market value for some of their competitors. To be fair to the makers of the visual, the design gaffe lands up inflating JP Morgan's losses as well but not as much as some of the others. Few laments that this is a common design flaw which is overlooked by many graphic designers.

Encoding quantitative values as areas of circles is of course more accurate than diameter encoding but still leaves us with entities which are difficult to compare. William Cleveland in his seminal book *The Elements of Graphing Data* clearly illustrates that it is difficult for us to estimate differences in area versus differences in length.

In my data visualization classes, I often find myself trying to convince the participants about this particular perceptual element of human cognition. To drive home this point, I typically conduct a quick exercise adapted from Stephen Few's article on pie charts.[7] The participants are shown two circles of different sizes and asked to estimate the area of the larger circle if the smaller one's area is assumed to be 1. The participants' estimates for the size of the larger circle usually range from about 5 times larger to 36 times larger and in rare instances up to 64 times larger. The two circle areas actually have a ratio of 1:16 and only a couple of participants usually manage to get the correct ratio.

---

articles/visual_business_intelligence/our_fascination_with_all_things_circular.pdf

[7] Stephen Few, 'Save the Pies for Dessert', *Visual Business Intelligence Newsletter* (August 2007), https://www.perceptualedge.com/articles/visual_business_intelligence/save_the_pies_for_dessert.pdf

This simple exercise with its wide range of answers instantly reveals to the class how inept our human brain is in comparing areas. After this exercise, most participants are convinced about the disadvantages of 2D encoding constructs versus 1D constructs like bar charts. I hope this discussion has also convinced the reader that all kinds of 2D constructs including pie charts, area graphs, bubble charts and Marimekko charts (tree maps) should be used with caution and only for specific scenarios where 1D alternatives such as bars, lines and dots are not viable.

This conclusion begets the following question: If our brains are not very accurate at estimating areas, how good or bad are we at estimating volumes? The short answer is that we are much worse at estimating volumes than areas! For a slightly longer discussion on volume encoding or in other words 3D constructs in graphs, turn over to the next chapter.

CHAPTER 3

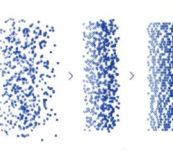

THE DECEPTION OF
THE THIRD DIMENSION

There is only one place for most 3D graphs, and that is the trash bin! The large majority of 3D graphs used in general or business context are confusing to interpret and prone to be misread. Popular data visualization tools such as ggplot for R and Tableau do not even have the provision for 3D graphs, since most data visualization experts discourage their usage.

Tableau and ggplot are based on the grammar of graphics,[1] a layered framework to concisely describe the components of any type of graph. The grammar of graphics gained popularity via ggplot and over time became the substrate for other visual analysis tools such as plotline for Python, D3 and Tableau. It has over time become a robust standard for visualizing data and composing well-designed graphs based not on their names but on statistics and primary visual elements—and it does not allow 3D graphs.

Why are 3D constructs so frowned upon by most data visualization experts? Representing quantities with 3D objects poses multiple problems. Edward Tufte in his book *The Visual*

---

[1] Leland Wilkinson, *The Grammar of Graphics* (Statistics and Computing) (Heidelberg: Springer–Verlag, July 2005).

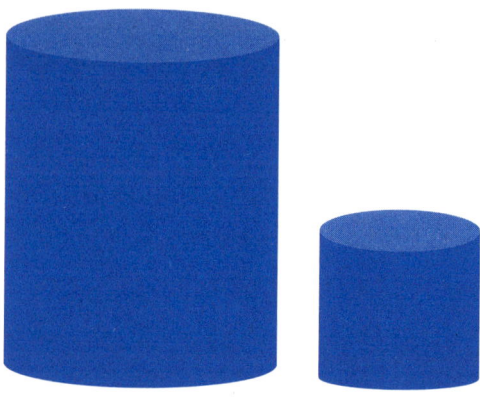

Figure 3.1: Using 3D objects for values, if the smaller barrel represents the value 10, what is your estimate for the value represented by the bigger barrel?

Display of Quantitative Information illustrates this with a graph which uses barrels of oil to represent the price of crude oil over time. Look at Figure 3.1, which contains two barrels. Can you estimate the amount represented by the bigger barrel if the smaller barrel represents 10 units of the currency? Do write down your answer before reading further.

What did you compare—heights, surface areas or volumes of the two barrels? If heights are compared, then the bigger barrel represents 20 units, and if we compare the surface areas, it represents around 30 units. If, however, the volumes are compared, the bigger barrel is more than five times the size of the smaller barrel and hence represents around 50 units, more than double what the height represents! How close did you get? Most viewers tend to compare volumes, which are, however, difficult to estimate accurately.

The conundrum in Tufte's example is similar—while the prices were represented by the height of the barrels, the graph was misleading because most viewers compared the volumes of the barrels.

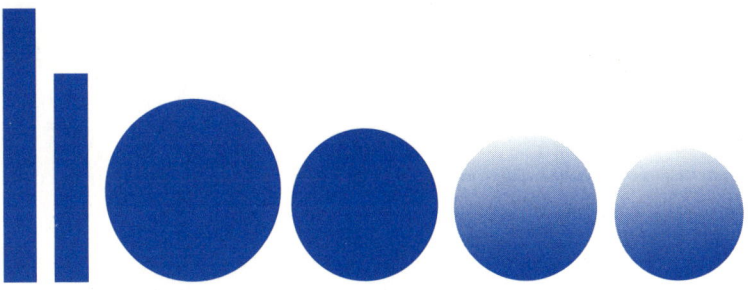

Figure 3.2: Comparing size of objects in 1D, 2D and 3D.

William Cleveland and Robert McGill, in their landmark paper on graphical perception,[2] establish through a series of experiments that visually judging differences in quantities is easiest when the quantities are represented in one dimension (e.g., as heights of bars) instead of areas of shapes or volumes of objects. As we discussed in the earlier chapter, representing quantities as 2D shapes, for example, as areas of circles, makes it much trickier to estimate differences in quantities. This problem gets further heightened when we need to estimate quantities represented as 3D objects, for example, as volumes of spheres.

Try this quick exercise out for yourself. In Figure 3.2, we have pairs of bars, circles and spheres. Can you estimate the ratio of the heights of the two bars? Now estimate the ratio of the areas of the two circles. Now move on to the spheres.

If you guessed that the smaller bar is around three-quarters of the bigger bar, you are absolutely right! What were your estimates

---

[2] William S. Cleveland and Robert McGill, 'Graphical Perception: Theory, Experimentation, and Application to the Development of Graphical Methods', *Journal of the American Statistical Association* 79, no. 387 (September 1984): 531–554.

for the circles and spheres? Are you surprised to know that their ratios are also 4:3? The volume of the smaller sphere is indeed three-quarters the volume of the bigger sphere. You might have realized through this exercise that it is certainly very difficult for our brain to estimate differences in volume. In the study mentioned earlier, Cleveland and McGill establish that viewers tend to underestimate areas and volumes. In fact, they also show that the underestimation in volumes (3D objects) is more acute than in areas (2D shapes).

Given how human perception and cognition work, it seems prudent to refrain from using 3D objects to represent values, especially when there are more dependable alternatives available, like the length of a bar or the height of a dot.

There are primarily two scenarios where 3D constructs are used in business charts. The first scenario pertains to situations where there are three variables which need to be presented. As an example, consider a company which wants to compare monthly sales for three different product categories in four different regions, via three different channels. With one axis for the quantitative value (sales in this case), the remaining two axes could be used for two of the three variables, say region and channel. The third variable could be accommodated by using pairs of clustered bars (we will come back to this scenario in a bit).

The second scenario where 3D graphs are used, typically involves only one variable which needs to be visualized. Nevertheless, a 3D object or space is used in such situations primarily to add some 'visual appeal' (for lack of a better way to articulate the reason!) to the graph. The oil barrel example discussed earlier falls under this category. Rest assured that for both these scenarios, there are more effective ways to depict the data than to resort to 3D constructs, as we shall shortly see. We will first tackle the second scenario where 3D objects are employed to make the graph visually appealing and then go on to the first scenario of visualizing multiple variables.

Do note that this discussion does not pertain to the use of 3D constructs in certain scientific studies where sometimes there is a genuine need to visualize a 3D space.

USING 3D OBJECTS FOR VISUAL APPEAL

All objects including spheres, cones, towers and 3D pies are completely unnecessary for representing data and should be discarded outright. Figure 3.3 depicts the use of some of these objects as used in data visualizations (left), juxtaposed with a more accurate way to depict the same data (right). Notice how in all four cases, using 3D objects makes the data much more difficult to interpret.

In the first example, planes are used instead of lines to show quarterly trends for two products. The perspective used in the 3D construct would make us believe that Product 2's sales are always higher than those for Product 1. However, this is only because of the angle at which this graph is viewed, and removing the 3D perspective and using a simple line chart shows us that Product 2's sales are actually lesser than Product 1's sales in the first quarter.

Moving on to the towers in the second row of Figure 3.3, a quick look would suggest that quarter 1 sales are a little more than $60,000 and quarter 4 sales are slightly above $40,000. Again, these are only illusions and removing the 3D element and using simple bars shows us that sales in the first quarter add up to exactly $60,000 and quarter 4 sales are actually less than $40,000. How could the tower constructs be so misleading and why are we not able to interpret them correctly? The fault, dear

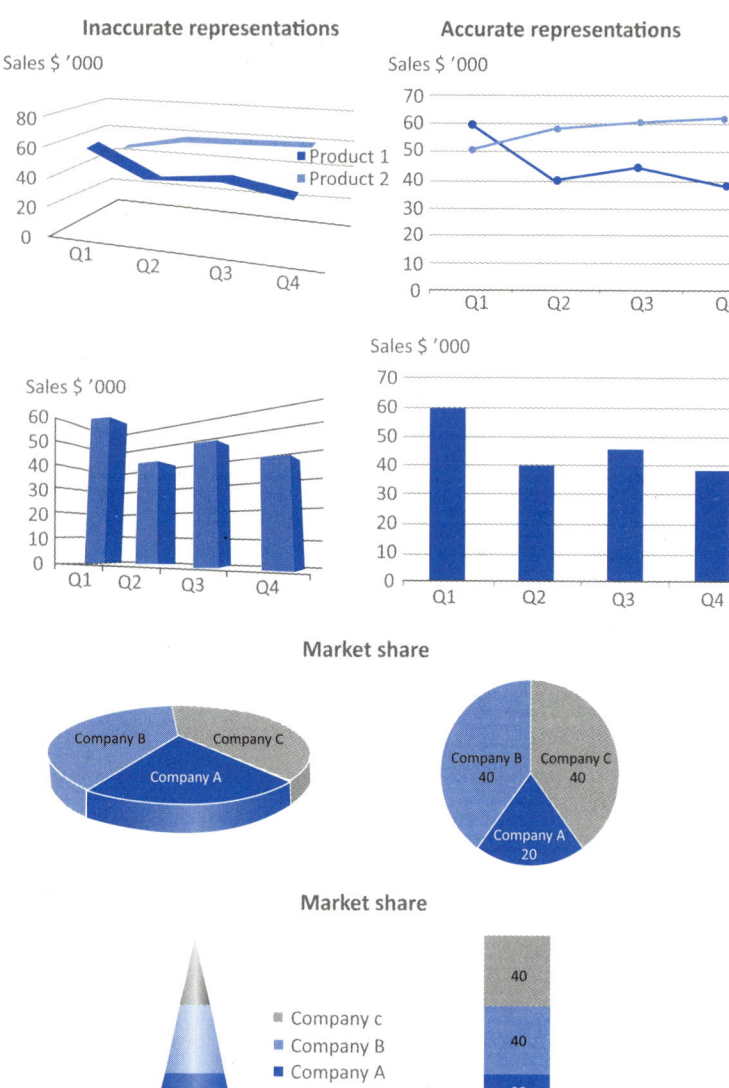

Figure 3.3: Unnecessary 3D constructs in graphs (left) and their more accurate counterparts (right).

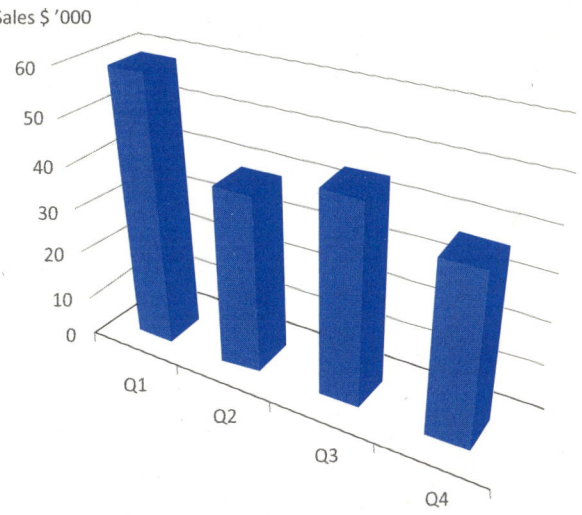

Figure 3.4: The towers do not touch the back wall, necessitating the extrapolation of the roof of a tower to the back wall to judge the correct height of each tower.

reader, is not in your data interpretation skills. Look carefully at the depth of the towers. You will notice that the towers stop short of the back wall in the graph. Figure 3.4 provides yet another perspective of the same graph in which this is clearly visible. To estimate the height of the tower correctly, the roof of the tower needs to be extrapolated backwards to touch the wall. That's a lot of work to interpret a single data point accurately! If by this time you are a bit dizzy with all the tracking of planes and lines this example has required, do yourself and your audience a favour in future and use a simple bar chart.

3D pie charts can be even more misleading than other 3D constructs discussed so far. From the pie chart on the left in Figure 3.3, try to estimate Company A's market share. This is an exercise I often use in my classes and most participants estimate that Company A owns a third of the market share. They are often shocked when the corresponding 2D pie chart (on the right in the same figure) reveals that Company A owns only

20 per cent of the market share. The subterfuge in 3D pies is easy to achieve. Slices placed closest to the viewer appear bigger than they actually are and slices placed at the back appear smaller than their actual size. As long as the category that needs to be inflated is positioned closest to the viewer, the effect is remarkable! In fact, with 3D pie charts, the creator has full control of the amount of distortion to be introduced in the graph by breezily changing the 3D rotation angle of the pie. The lower the perspective, the larger the front category looks.

In *The Visual Display of Quantitative Information,* Tufte introduces the concept of the 'lie factor' which defines the degree to which a graph is untruthful. The lie factor is defined as the ratio of the size of an effect in the graph to the actual size of the effect in the data. A graph with no distortion should have a lie factor close to one. A lie factor greater than one implies that something is being overstated and a lie factor less than one would mean that something is being understated.

With respect to the 3D pie chart in Figure 3.3, if a viewer esti- mates Company A's share as 30 per cent instead of 20 per cent, the lie factor is 30/20 or 1.5. This means that the effect appears approximately 50 per cent bigger than it actually is. By changing the angle of viewing of the pie with a couple of clicks, this lie factor can easily be increased or decreased. Figure 3.5 shows the same 3D pie chart at different angles of perspective. All three categories are equal and represent a third of the pie, yet note how Category B appears to increase with the change in perspective.

If you have encountered 3D pies before, take a moment to think about how the different categories were relatively placed and if perhaps the advantageous position of the front slice was leveraged to inflate a chosen category. If you have not encountered 3D pies and think that this is rather an obvious trick that nobody would dare play, let it be known that they are

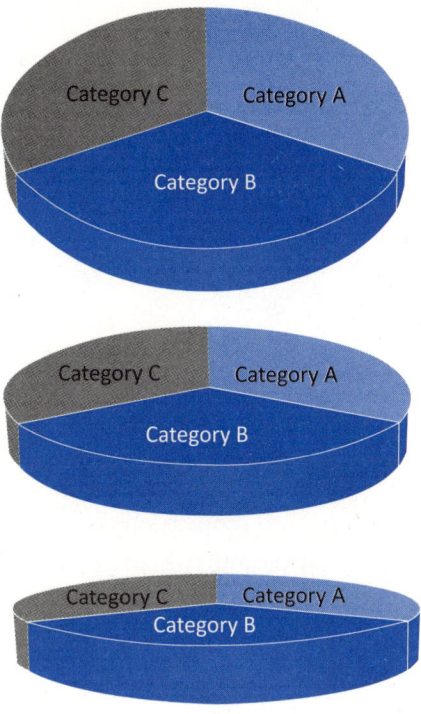

Figure 3.5: Changing the perspective of the 3D pie is all that is needed to make the front slice (Category B) appear bigger and bigger.

quite ubiquitous. Prestigious companies like Apple have been caught using the 3D pie to their advantage. In a keynote address, Steve Jobs used a 3D pie chart to illustrate the market share of Apple and its competitors.[3] A recreation of the original 3D pie and a more accurate redesign using a normal pie chart is shown in Figure 3.6. It's subtly done, but notice how Apple's slice at 19.5 per cent actually looks bigger than the 'Other' category, which is 21.2 per cent.

---

[3] https://www.engadget.com/2008-01-15-live-from-macworld-2008-steve-jobs-keynote.html

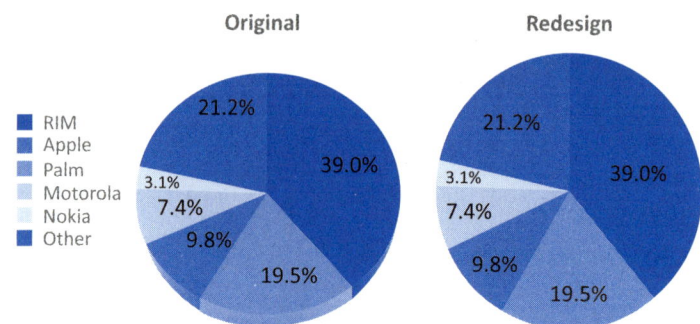

Figure 3.6: A recreation of the 3D pie chart of market share distribution used by Apple (left) and a more accurate 2D redesign (right).

The last row in Figure 3.3 uses a cone to represent the market share of three companies (constructed in MS Excel). It is straightway evident that despite a market share of 40 per cent, Company C's share appears much smaller than that of Company B, which is also 40 per cent. In fact, Company C's share looks almost the same as Company A's share, which is actually half the quantity.

By now, the astute reader must have guessed what is going on. The software uses the height of the various sections of the cone to represent quantities. The human eye of course perceives an object and compares the volume of the different sections. Since the bottom section is much wider than the apex, this cone becomes a rather nonsensical representation of the data. The stacked bar on the right is a more accurate way to represent the same data. Recall Tufte's principle of amount of data ink used to represent quantities, which was discussed in Chapter 2. in the stacked bar, the quantity of ink used to represent each category is proportional to the numerical quantity it represents, whereas in the cone, the principal of proportional ink is violated to an unacceptable degree.

Hopefully, the four examples discussed in this section have convinced the reader that 3D constructs are completely unnecessary when the data has only one or two dimensions (sometimes called variables). These examples also illustrate how 3D constructs are not only superfluous but sometimes outright dangerous, with the potential to create grossly misleading data visualizations.

But what about situations where there are indeed more than two variables to capture on a 2D space like a page of a report or a computer screen. Is it then justifiable to use a 3D graph? The next section deals with this dilemma.

THIRD DIMENSION TO INCORPORATE A VARIABLE

Let us revisit the example introduced in the previous section where data across three variables needed to be represented— sales in four different regions via three different channels and three different product categories.

Instead of three variables, let us first tackle a simpler scenario with only two variables, regions and channels. Our graph needs to display total sales figures for four different regions and three different channels. Figure 3.7 depicts three different possibilities for this scenario—a 3D tower construct, a clustered bar chart or a stacked bar chart.

With the tower construct, not only is it difficult to compare the heights of any two towers, but some of the towers are also only partially visible. In fact, the tower representing sales via Channel 3 in Region 1 is completely hidden behind the other towers. Deciphering heights of individual towers is an equally daunting task. Can you find out sales for Region 4 under

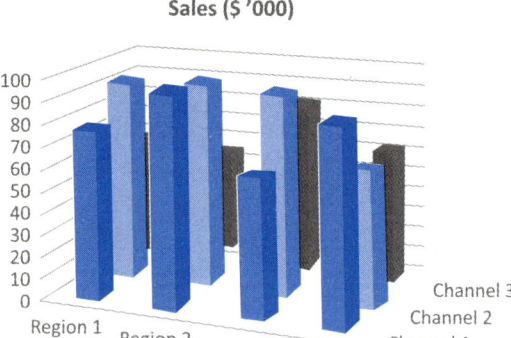

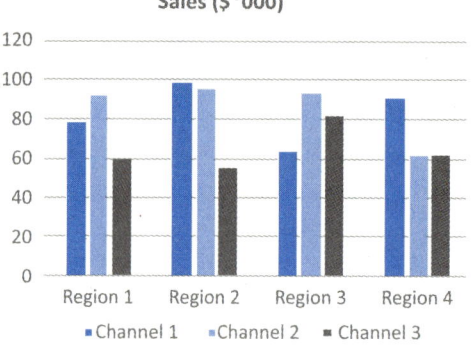

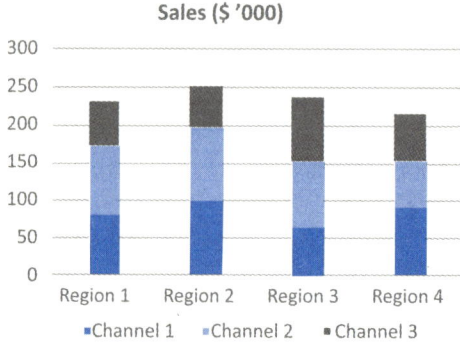

Figure 3.7: Three different options for displaying sales figures for two variables—regions and channels.

Channel 3? It requires considerable effort to trace the corresponding grid line from the back wall to the side wall and then the y-axis. The correct sales figure is 60. Notice how easy it is to answer the same question in the clustered bar graph (second graph in Figure 3.7). With all the bars clearly visible in the 2D construct, the guesswork involved in interpreting the numbers and comparing relative heights is eliminated.

What about the third option, the stacked bar chart? While the stacked bar construct is definitely more compact than the other two constructs, estimating individual values or comparing two values is tricky. It is easy to interpret values for the bottom category (Channel 1) as all the bars have a common baseline, but it is far trickier to compare across channels or within the other two channels.

Try to answer the following question using first the stacked bar chart, then the 3D tower chart and finally the clustered bar chart.

Among Regions 1 and 3, which one has more sales under Channel 2.

The correct answer is Region 3, but as you would have realized, it is next to impossible to deduce that from either the 3D tower chart or the stacked bar chart. The clustered bar chart with a common baseline for all the bars allows for much more accurate comparisons. Cleveland and McGill established this fact fairly early in their 1985 article in *Science* magazine[4] when they reported that 'humans are fairly good at comparing differences in length, but only when things share a common reference point.'

---

[4] William S. Cleveland and Robert McGill, 'Graphical Perception and Graphical Methods for Analyzing Scientific Data', *Science* 229, no. 4716 (1985): 828–833.

Now that we have established the construct that works best for two variables, it is time to revisit the original scenario with three variables: region, channel and product category. Adding additional towers to our first option will only make it more unreadable, and adding a third dimension to either of the two bar charts (and thus converting them to towers) will create the same set of data interpretation problems associated with tower constructs which were already discussed earlier.

The path out of this perplexing situation is to use what are called 'small multiples'. The small multiples construct (also called a trellis chart) was proposed by Tufte in *The Visual Display of Quantitative Information* as a solution to effectively display multivariate data. Applying the small multiples construct to our example, each product category could get its own graph as shown in Figure 3.8.

Although there are three separate graphs used here, they are placed together in one view to enable comparisons across graphs. A quick look at the graphs reveals that across all regions and product categories, sales via Channel 3 are almost always lower than the other two channels. The graphs also tell us that sales figures for Product Category 3 are noticeably lesser than the other two categories.

Tufte describes small multiples as resembling 'the frames of a movie: a series of graphics, showing the same combination of variables, indexed by changes in another variable'.

In our example, the index variable is the product category, with everything else in each graph—the quantitative axis, primary and secondary category order and colours and grid lines—remaining exactly the same. All the design elements in each entity of the small multiples construct are identical,

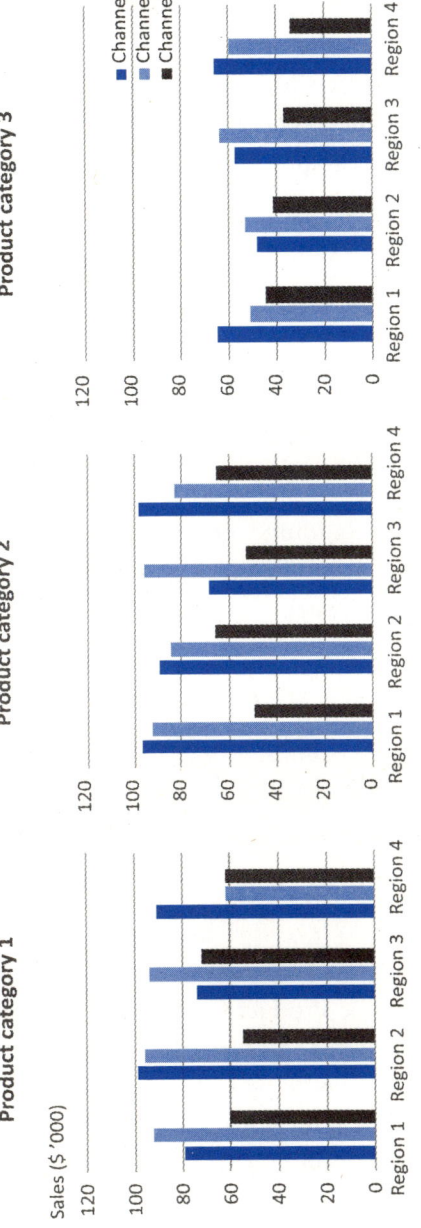

Figure 3.8: Representing sales for three variables—region, channel and product category—using the small multiples construct.

making it easier for the viewer to interpret the large amount of data presented across the graphs. The viewer needs to understand the layout of just the first graph and then use that understanding to interpret only data variations in the other graphs. Note how in Figure 3.8 a common legend and a common y-axis label for all the graphs binds the three graphs into one visual entity, encouraging the viewer to compare data across the graphs.

Small multiples can be powerful constructs for analysing multivariate data, provided the design of each graph in a collection stays exactly the same. Some misleading graphs are created wilfully by humans and others are unwittingly created by the machine itself. When creating small multiples, certain software packages (including MS Excel) automatically adjust the quantitative axis range according to the range of the data. This feature can be disastrous for the small multiples construct by creating a very deceptive visual.

Figure 3.9 depicts a scenario where the scale of the third graph has been automatically adjusted by the software to better capture the range of the data in that particular graph.

Do you see the problem? Sales for Product Category 3 now look better than the other two categories, except that with the changed y-axis in the third graph (the y-axis only goes up to 70 instead of 120), an inter-graph comparison of the lengths of the bars is meaningless.

On the same lines, suppose Product Category 2 was not sold in Region 3. The correct way to design the second graph of the small multiples construct would be to leave a gap for Region 3 (right side of Figure 3.10), rather than skipping it altogether and replacing that space with Region 4 (as shown on the left side of

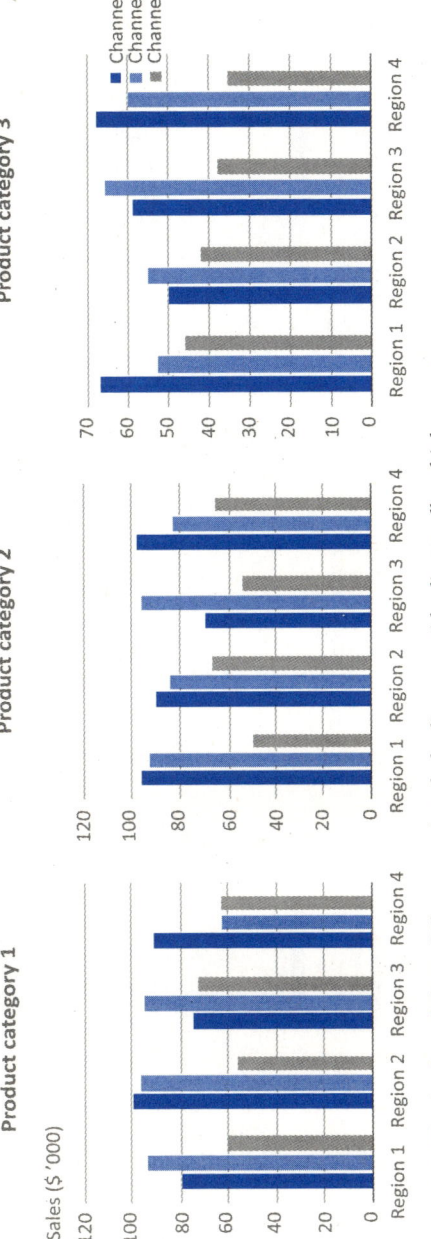

Figure 3.9: The third graph has a different y-axis scale, leading to a misleading small multiples construct.

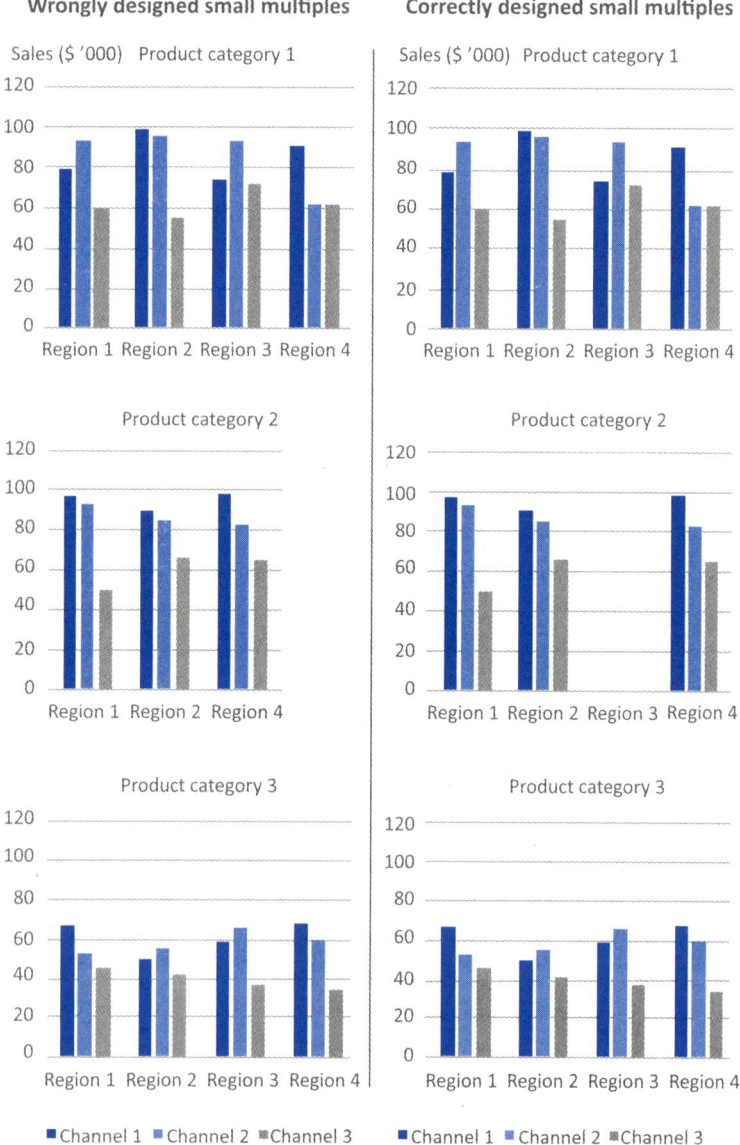

Figure 3.10: Consistency in placement of categories is an important aspect of designing small multiples.

the same figure). The danger with replacing the space with Region 4 is that it can easily be misconstrued as sales figures for Region 3, leading to incorrect conclusions.

Therefore, when creating small multiples, make doubly sure that the AI in your software is not outthinking you and misleading your viewer. Of course, some small multiples seem skilfully constructed to misrepresent. Figure 3.11 shows a recreation of the graphs depicting revenue and cost figures in Zomato's 2019 annual report[5] and a more accurate redesign (first discussed in an article in The Ken[6]).

In the small multiples construct used by Zomato, revenue and cost figures for two years are placed next to each other, encouraging a comparison between the two graphs. Note that the quantitative axis is conveniently missing. The viewer may compare the lengths of the revenue and cost bars for 2018 and assume that the company had made a profit. Similarly, a comparison of the lengths of the 2019 bars alludes to a break-even performance. The catch is that the two graphs have very different y-axis scales. The y-axis of the revenue graph goes up to $250 million, while the y-axis of the cost graph goes up to $600 million.

The redesign (with both axes going up to $600 million) paints a more accurate picture. It is now evident that for both years, total cost was higher than revenue. In fact, the losses in 2019 seem quite alarming. Is this an instance of wilful

---

[5] https://www.zomato.com/blog/wp-content/uploads/2019/04/Zomato%E2%80%93Annual%E2%80%93Report%E2%80%93FY19.pdf
[6] Sumanth Raghavendra, 'Zomato Faces Dark Clouds with a Gold Lining' (2019), https://the-ken.com/story/zomato-gold-report/?utm_source=daily_story&utm_medium=email&utm_campaign=daily_newsletter

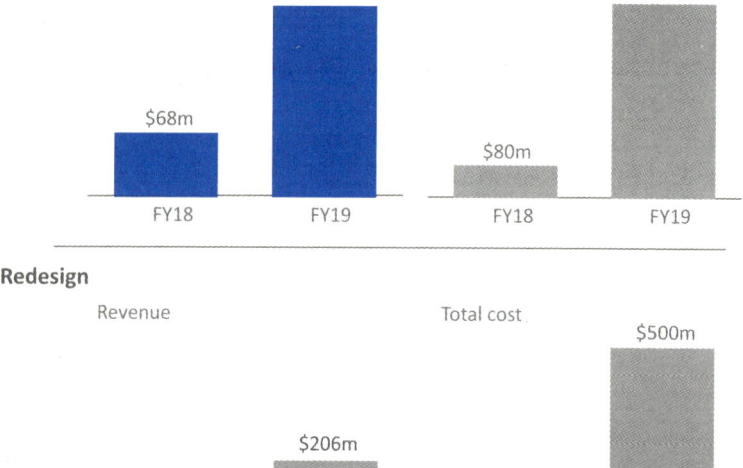

Figure 3.11: Revenue and cost figures as reported in Zomato's annual report (top) and a more accurate redesign (bottom).

misleading or a careless oversight by the creator of the visual? You, the astute reader, can decide, but since the actual profits are not directly stated anywhere in the report, the latter seems less probable.

In his book *Show Me the Numbers*, Stephen Few proposes that we pay attention to the following four elements while constructing small multiples: consistency, arrangement, sequencing and grid lines.

Consistency: As mentioned earlier, each graph in a small multiples construct should be consistent in design. Elements such as font type, colour choices, size, aspect ratio, axis range, order and placement of categories should be the same. Any divergence could make a comparison across graphs confusing or misleading.

Arrangement: When arranging the graphs, a vertical versus a horizontal arrangement could aid different types of comparisons. Look at the horizontal arrangement in Figure 3.8 versus the vertical arrangement in Figure 3.10 (right). It is easier to compare how a particular channel is doing in a particular region across product categories if we use the vertical construct. By contrast, an overall comparison across product categories is easier in the horizontal construct.

Sequencing: When possible, sequence or order graphs in increasing or decreasing values for more meaningful comparisons. In Figure 3.8, Product Category 3 has the least sales and hence is placed last so that Product Categories 1 and 2, which have closer values, can be compared more accurately.

Grid lines: If the small multiple construct uses a grid or trellis form, the user may be confused about whether the page should be scanned vertically or horizontally. Figure 3.12 illustrates how grid lines can be used to guide the user to either compare the graphs horizontally or vertically. In design parlance, these small details are defined as the affordances which can be built into a product to provide suggestions for how it should be used.

In fact, any well-designed object will provide cues to the user to operate it correctly. Think about the affordances which can be built into a door which opens only one way. An obvious way to ensure that a user knows when to push versus pull is to put up the respective signs near the handles. However, a little introspection will reveal that this solution may not work for all users. The signs will be useless for those who cannot read that language, for instance, international visitors not familiar with the

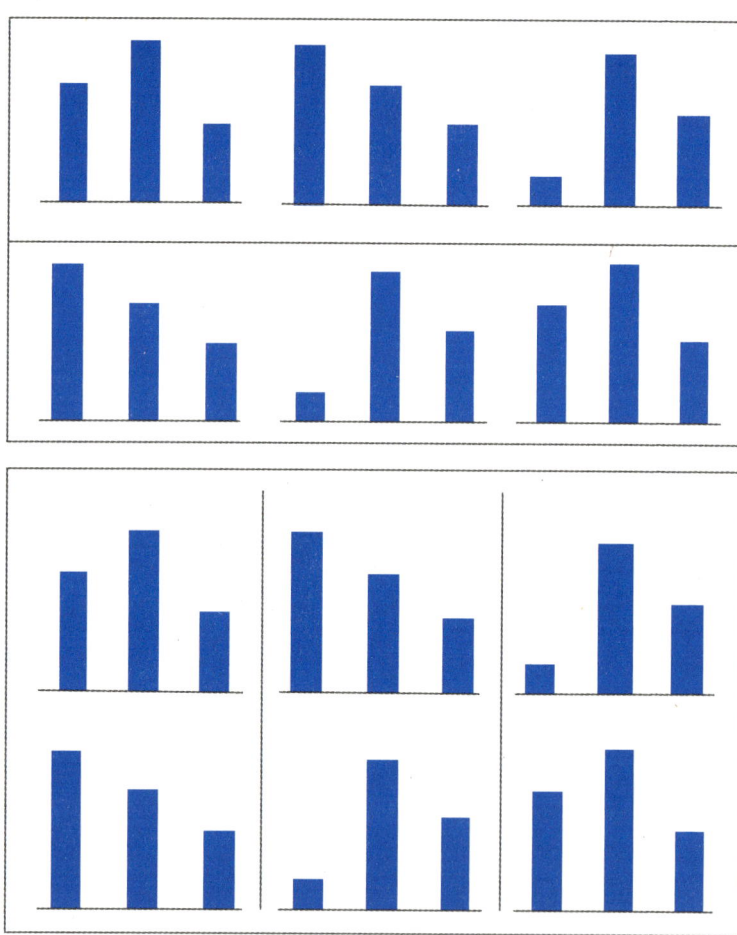

Figure 3.12: Horizontal (top figure) or vertical (bottom figure) grid lines in the small multiples construct provide cues about which graphs should be compared to each other.

language or small children. A better solution would be to place only a panel on the side which needs to be pushed and a handle on the other side.

While designing a graph, think about the affordances you can build into the visualization so that it is used in the way 'you' intend it to be used.

Hopefully, this chapter has convinced you that using 3D constructs to represent quantitative values creates visuals which are difficult to interpret and some which are downright deceptive. With better alternatives available such as simple lines, bars, stacked bars and well-designed small multiples, 3D objects are best discarded. To reiterate, the only appropriate place for all your 3D graphs is the trash bin, unless of course the aim is to confuse your audience.

CHAPTER 4

SPURIOUS TRENDS AND HOW TO SPOT THEM

Someone who needs to regularly present data, be it in reports, dashboards or presentations, is familiar with the holy grail of decisions: Should we use a table or a graph? Sometimes the answer is obvious, but many situations warrant a closer look at how the information presented is going to be used. In his book *Show Me the Numbers,* Stephen Few lays out a simple yet effective set of guidelines to help you make this decision. He suggests that if individual, precise values need to be looked up or specific values need to be compared, then tables work better than graphs. On the other hand, the strength of a graph lies in understanding the 'shape of the data' and 'spotting trends, patterns and exceptions'.[1]

However, using a graph to express a particular trend can be a double-edged sword. Without particular care to certain design elements, it is relatively easier to create a misleading pattern and lead the viewer to incorrect conclusions about the data.

Visual connections should reflect real connections.

Hadley Wickham,
Chief Scientist, RStudio

---

[1] Stephen Few, *Show Me the Numbers: Designing Tables and Graphs to Enlighten* (Burlingame: Analytics Press, 2012).

In this chapter, we investigate different ways in which graphs might visually convey a trend or connection not actually present in the data and then suggest alternate designs which reflect real connections. Our enquiry covers a range of topics, including the aspect ratio of a graph, unexpected change of scale, stacked area charts, effect of inverting the axis and finally dual axis graphs.

ASPECT RATIO OF GRAPHS

The aspect ratio of a graph is defined as the ratio of its width to its height. A square has an aspect ratio of 1:1, and a typical landscape photograph has a ratio of 3:2. Adjusting the aspect ratio of a graph can radically change the perception of a trend. Figure 4.1 shows the same graph with three different aspect ratios. Notice that all other design features including the y-axis scale are identical in all three graphs yet how they differ in the message conveyed!

Changing the aspect ratio in a graph changes the slope of the lines in a graph, hence leading to different possible conclusions about the same set of data. What then is the ideal aspect ratio for a given graph? While there have been a few studies on this particular question, the results are less than conclusive.

William Playfair, a data visualization pioneer who is attributed the invention of line, bar and pie charts during the turn of the 18th century, is known to have preferred an aspect ratio of 1.4:1 to 1.8:1.[2] Tufte and other experts recommend a common sense approach. They suggest that the nature of the data should decide the aspect ratio and that there is no single golden ratio which

---

[2] https://medium.com/nightingale/how-to-create-a-simple-yet-effective-bar-chart

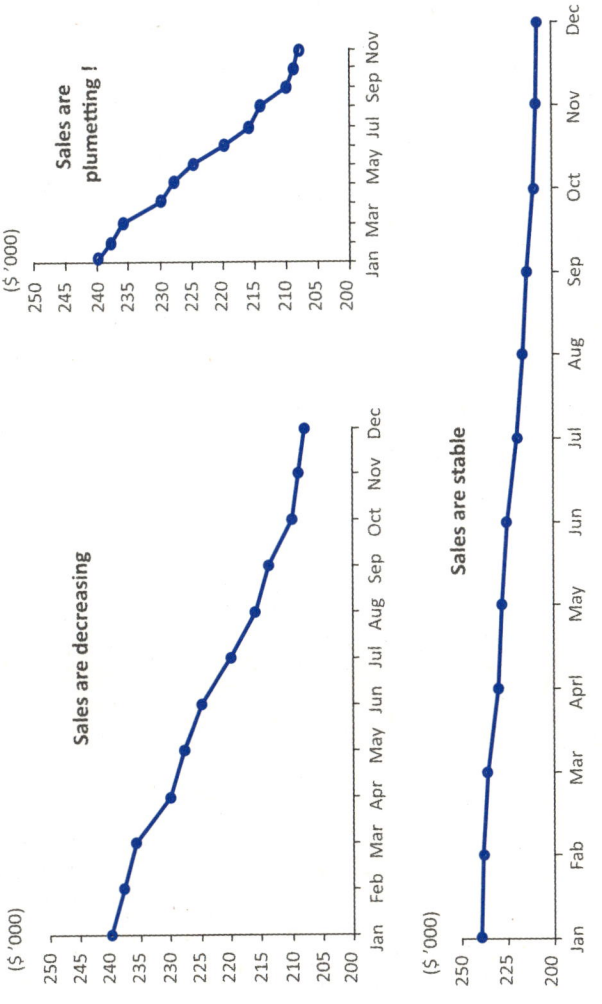

Figure 4.1: Three identical graphs with different aspect ratios.

can be adhered to. Tufte does mention that a default ratio of 1.5:1 (the standard 3:2 ratio for a landscape) can be followed if the data does not demand otherwise. Often while deciphering a trendline, the viewer needs to notice differences in the rates of change. Choosing an aspect ratio (more by trial and error than any golden ratio) which facilitates the comparison of proximal slopes is important in such cases.

In one of the few experimental studies on the topic of graph aspect ratios, Cleveland et al.[3] found that to facilitate accurate comparison of the slope of two lines in a graph, the average line slope should be 45 degrees. This concept, which has been named 'banking to 45 degrees' has become a de facto standard for the aspect ratio of line graphs. However, a more recent study by Talbot et al.[4] found that slope ratio errors are not minimized around 45 degrees and flatter graphs (i.e., shallower slopes) actually helped viewers make more accurate slope comparisons among pairs of lines.

Both studies, however, concur that slopes steeper than around 45 degrees make it more difficult to compare angles. Figure 4.2 contains three different aspect ratios for a graph, where monthly sales of two products need to be compared. Graph (a) is banked to 45 degrees, (b) has a steeper average slope and (c) is shallower. Judge for yourself if graphs (a) and (c) make it easier to compare the growth of the two products compared to graph (b).

---

[3] William S. Cleveland, Marylyn E. McGill, and Robert McGill, 'The Shape Parameter of a Two-Variable Graph', *Journal of the American Statistical Association* 83, no. 402 (1988): 289–300.

[4] Justin Talbot, John Gerth, and Pat Hanrahan, 'An Empirical Model of Slope Ratio Comparisons', *IEEE Transactions on Visualization and Computer Graphics* 18, no. 12 (2012): 2613–2620.

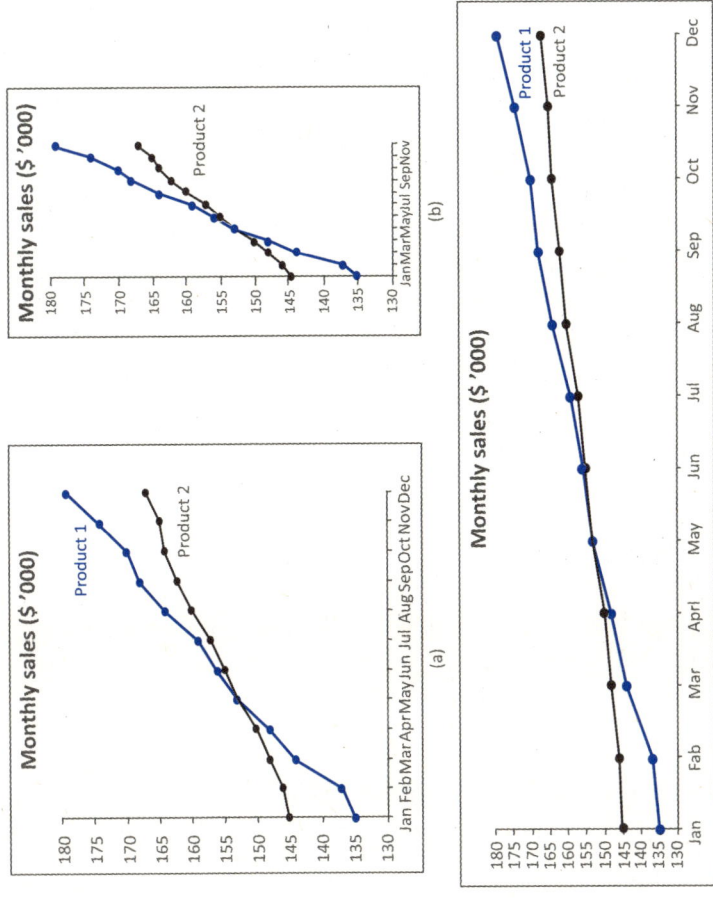

Figure 4.2: Comparing the slopes of two lines is easier when the average slope of the lines in banked to 45 degrees as in graph (a) or less as in graph (c). Steeper slopes (b) are more difficult to compare accurately.

With the range of confusing evidences discussed above, what is the takeaway for deciding the aspect ratio of a line graph, so that it helps the user understand the data correctly?

Extreme aspect ratios where the graph is very tall and narrow or broad and shallow should evoke some scepticism. Ask yourself if the data warrants such a distortion from the classic 3:2 ratio or there could be an ulterior motive for the oddly shaped graph.

Additionally, apart from comparing the slopes of lines, shallow slopes do not work well for other cases. Banking to 45 degrees as a general principle may still work reasonably well for many graphs. However, there might be valid reasons to deviate from this principle if the situation demands so, and a little trial and error will lead you to the right proportions.

While the 3:2 aspect ratio is a good starting point for most bar and line graphs, the same is not true for scatterplots. Since a scatterplot captures the relationship between two variables, a graph which is wider than higher may distort the correlation. Best practices indicate that scatterplots with an aspect ratio closer to 1:1 (a square) work well. The bottom line is that if you internalize Tufte's advice and 'avoid distorting what the data have to say', your graph will find itself a suitable aspect ratio.

UNEXPECTED CHANGE OF SCALE

In Chapter 2, we looked at possible issues if the quantitative scale of a bar chart does not start at zero or if there is a kink in the scale (broken axis). Our responsibility for creating well-designed scales does not end there! An equally problematic scenario is when the scale changes partway along the axis. While this situation is less common for a basic quantitative scale (most graph makes do not tamper with the uniform

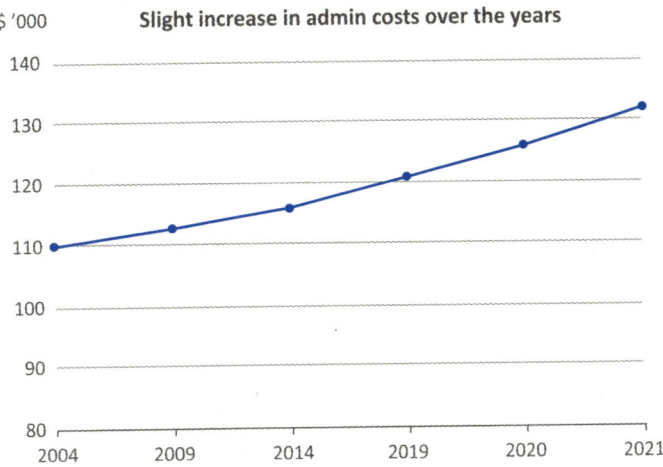

Figure 4.3: Uneven time scales can create a misleading trend.

jumps in numbers), it is quite prevalent when time scales or ranges are used.

Take a look at the example in Figure 4.3 which plots administration costs of a company over the years. Can you spot the problem here?

The trend of the line in the graph suggests a linear increase in administration expenses over a span of 18 years, from 2004–2021. However, a closer look reveals that the time scale on the x-axis is not linear. The data is initially plotted for every five years till 2019 but then changes to every year for the last three years. Figure 4.4 shows the correct way to plot this data with a consistent time gap on the x-axis scale.

By looking at yearly data, it is evident that there has been a sharp spike in administration costs in recent years, a fact which was entirely hidden in the previous graph because of the uneven time scale. The first graph plotted costs after a gap of five years and then without warning changed the scale to every year from

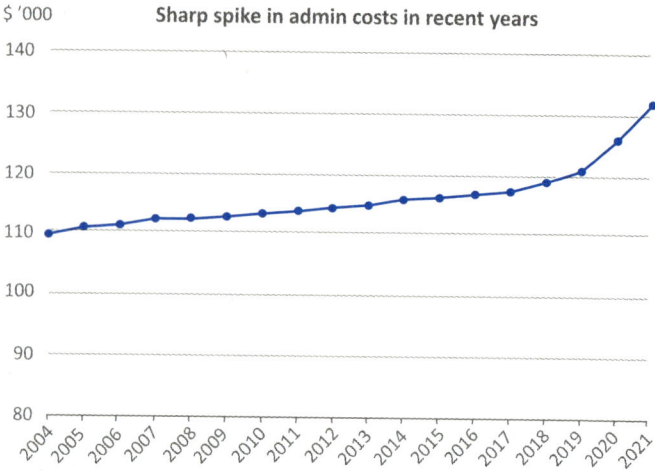

Figure 4.4: A consistent time scale reveals the actual trend in the data.

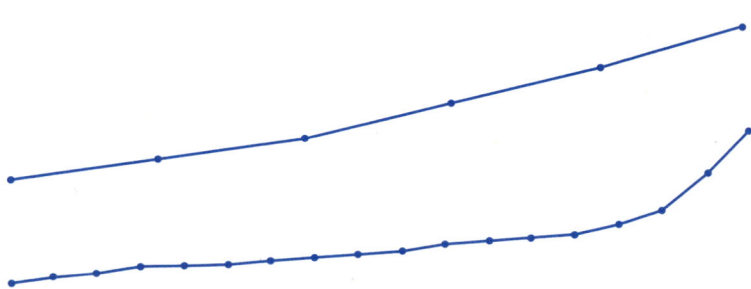

Figure 4.5: Both lines from the earlier charts are placed together to accentuate their difference in shapes.

2019 onwards. This ploy hid the sudden increase in costs in 2020 and 2021. Figure 4.5 shows both the trend lines together to accentuate the difference.

Sometimes data for all the time stamps may not be available, and hence plotting all the points may not be a possibility. In our example, let us suppose that the only data points available were the ones used in the first graph. In such cases, the time

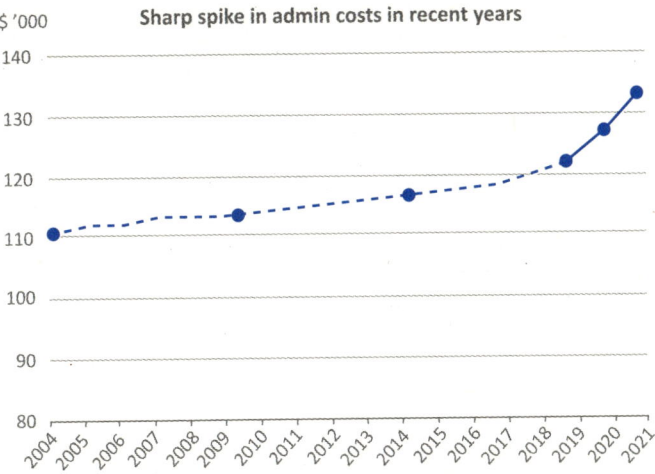

Figure 4.6: The correct way to plot a line if certain data points are unavailable—a uniform time scale plus dotted lines to indicate missing data.

gaps in the scale still need to be consistent and the line joining the data points should be converted to a dotted line to indicate the unavailability of data for certain timestamps (see Figure 4.6).

Another common situation where a change in scale can lead to a misleading trend is captured in Figure 4.7. Monthly sales figures are reported for six months, and there seems to be an alarming drop in June (graph on the left). Closer inspection reveals that there is in fact nothing to fret about. The data collection stopped in the middle of June, and the sales figures reported for June represent the number of units sold in the first 15 days of the month. The redesigned graph on the right captures the correct trend by indicating both the actual sales so far and the projected value for the whole of June.

Note the usage of dotted lines in both this and the previous example to clearly demarcate extrapolations/interpolations from known values. Dotted lines in data graphs have a sematic

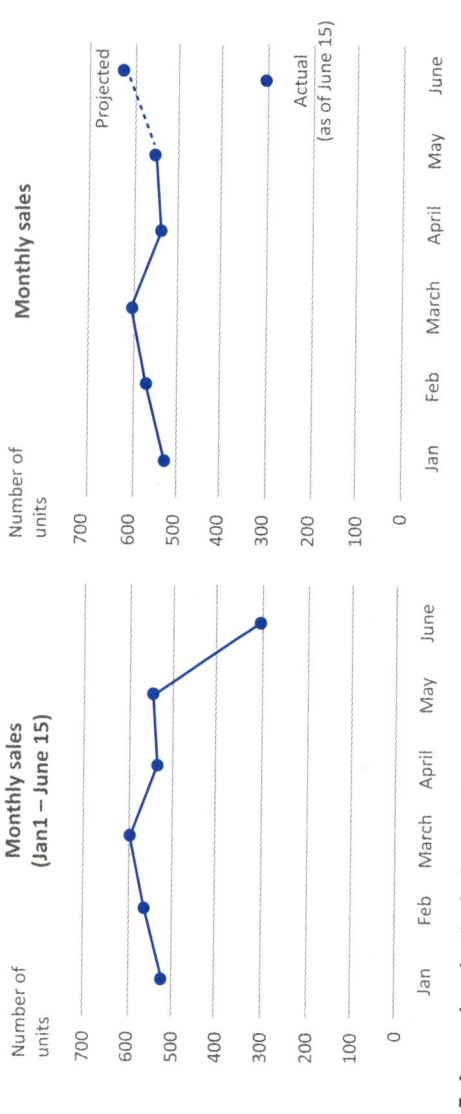

Figure 4.7: Incomplete data in the last month creates a misleading trend (left) which can be corrected by plotting both actual and projected numbers (right).

attached to them—they caution the viewer about predictions or projections. Hence, they should be used only for similar situations and never as a substitute for line colour or marker shape, which are commonly used to represent different categories.

We have now covered enough ground to revisit one of the examples of misleading graphs which you were quizzed about in the Introduction (see Figure 4.8). If the answer had eluded you earlier, can you now figure out how this trendline might mislead the viewer?

The hint of course lies in the title of the graph which tells us that the data was compiled in early 2022, which means the profit numbers for 2022 and 2023 are predictions and not actual figures. The sharp upward spike in profits is only a conjecture (hopefully based on sound evidence and not wishful thinking) and should be clearly demarcated as such. The redesign in Figure 4.9 fixes the problem by using dotted lines to join

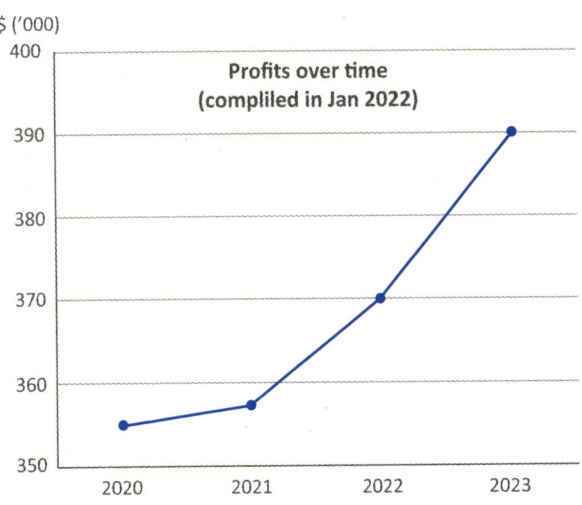

Figure 4.8: Can you spot what is wrong with this graph?

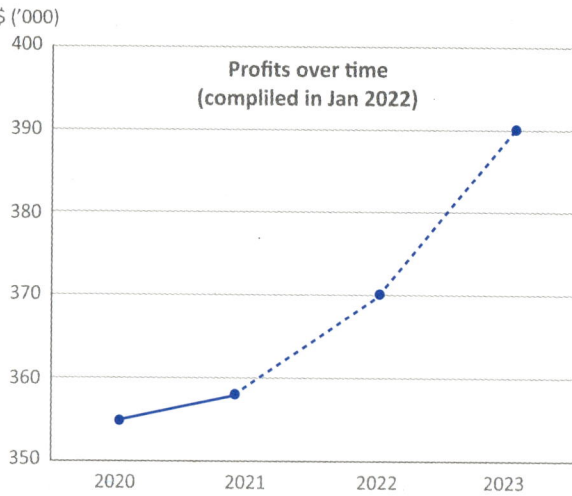

$ ('000)

Figure 4.9: Dotted lines should be used to clearly demarcate actual numbers from projections.

projected values, thus clearly communicating which part of the trendline is based on actual profits and which part is forecasted.

The examples discussed so far have dealt with time series data, but unexpected scale changes in other constructs can lead to equally muddled interpretations. Constructs such as histograms and percentile graphs which group data into ordered bins also need utmost care with their axis design.

Examine the histograms in Figure 4.10, which shows the distribution of employees' ages in two different companies. From the shape of the data, one could conclude that Company 1 has a larger proportion of older employees, since the graph is heavier on the right. In contrast, Company 2 has what appears to be a normal distribution with most employees in the middle age group of 40–45 years, with the numbers tapering off symmetrically on both sides.

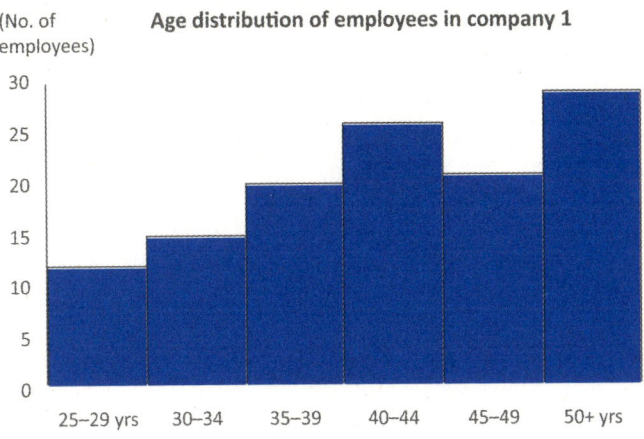

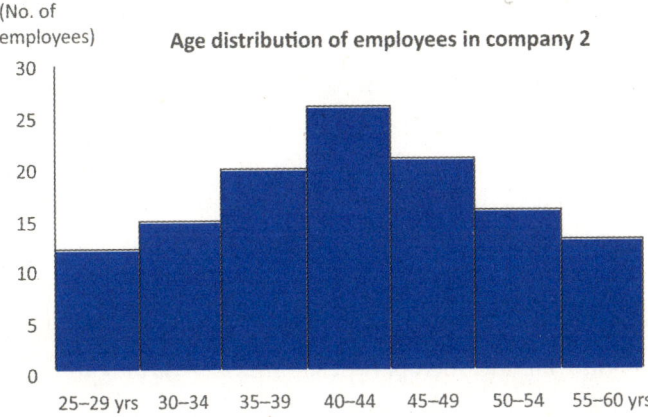

Figure 4.10: Which company has more older employees?

Does it surprise you then to know that the underlying data in both the graphs is exactly the same. Can you spot the problem?

The same oversight of an uneven scale is the issue in this example as well. The histogram for Company 1 has a bin width of 5 years for all the bins except the last one which has a bin width of 10 years. Essentially, the last bin (50+ years) is actually two bins (50–54 years and 55–60 years) combined as one, which irrevocably changes the shape of the data. The histogram for Company 2 with uniform bins is the accurate version and allows the viewer to discern the actual shape of the distribution.

While this particular lapse of unequal bin widths is quite common to histograms and frequency polygons, it is not the only concern to watch out for while using these constructs. Visualizing the distribution of a dataset entails that a continuous range of data (the age range of employees in this case) is chopped up into bins. The visual then represents the count of occurrences (or frequency) in each bin. While creating a histogram to plot and understand the nature of a distribution, it is important to experiment with a variety of uniform bin widths, as illustrated by the next example.

The four histograms in Figure 4.11 are based on the same dataset but notice how the interpretation of the nature of the distribution varies as the bin width is decreased (or, in other words, the number of bins is increased). What looks like a normal distribution when 4 bins are used turns out to be a bimodal (2 peaks) distribution when the number of bins are increased to 20!

Useful and accurate histograms not only have bins of the same data width but also the appropriate number of bins to tease out the true shape of the distribution.

STACKED AREA GRAPHS

A construct which is popular but particularly problematic is the stacked area graph, which typically shows how the quantities of a certain variable have changed over time for different categories. Figure 4.12 is the stacked area graph from the introductory quiz in the first chapter which shows expenses over the last six months for four departments. Were you able to spot the problem here?

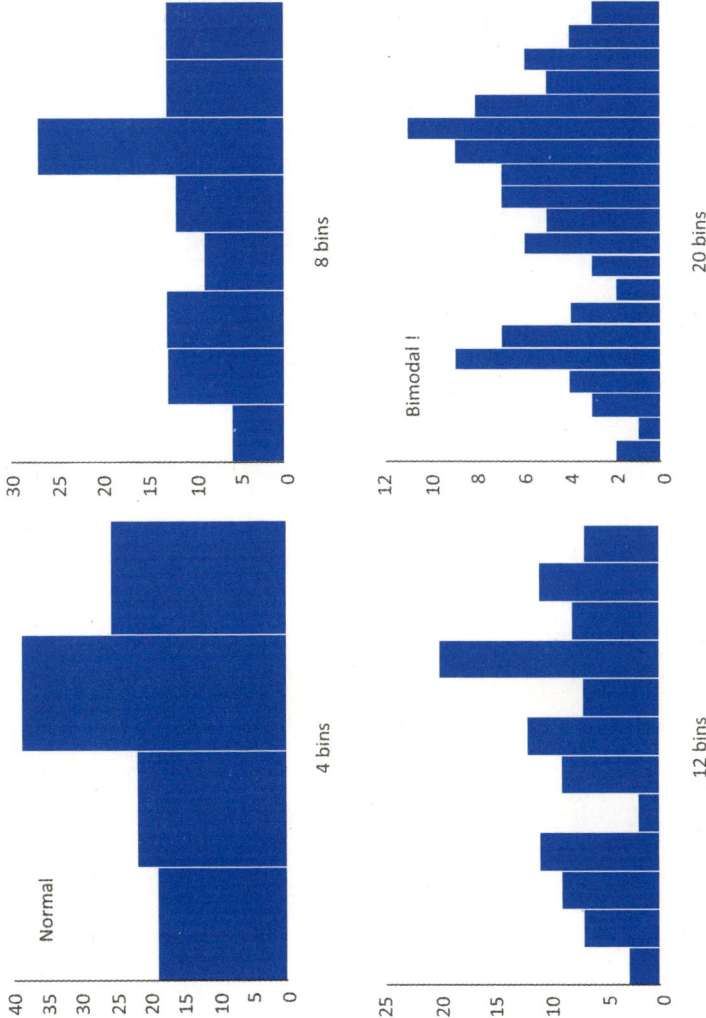

Figure 4.11: While creating histograms, try out different bin widths to tease out the actual shape of the distribution.

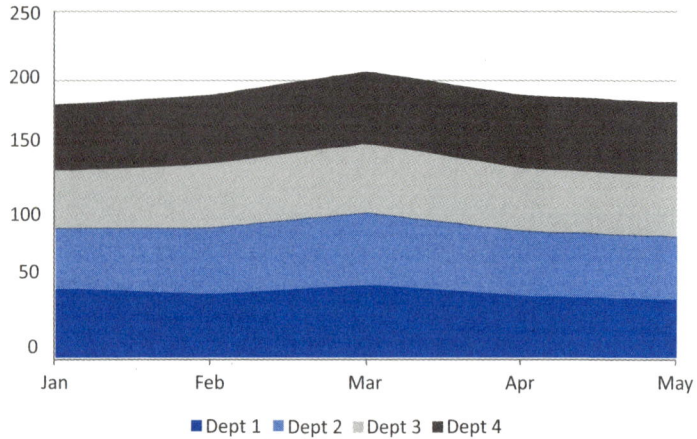

Figure 4.12: Department-wise monthly expenses as a stacked area graph.

A quick glance at the graph leads us to infer that except for a slight peak in March, all four departments have incurred steady expenses across the six months covered in the dataset. This interpretation of the expenses is actually completely incorrect as revealed by an alternate construct to depict the same data, a simple line for the expenses of each department (Figure 4.12).

The line graph in Figure 4.13 reveals that while other departments' expenses have indeed peaked in March and then subsequently dropped, expenses for Department 2 have continued to increase across the six months to peak in May.

If you are wondering how this insight escaped you in the stacked area graph, do not feel disheartened. Stacked area graphs are notorious for hiding the actual nature of the data. Look at Figure 4.12 again. The categories are stacked upon each other, which means that the only category whose trendline is accurate is the bottom-most category (Department 1 in this case). It is

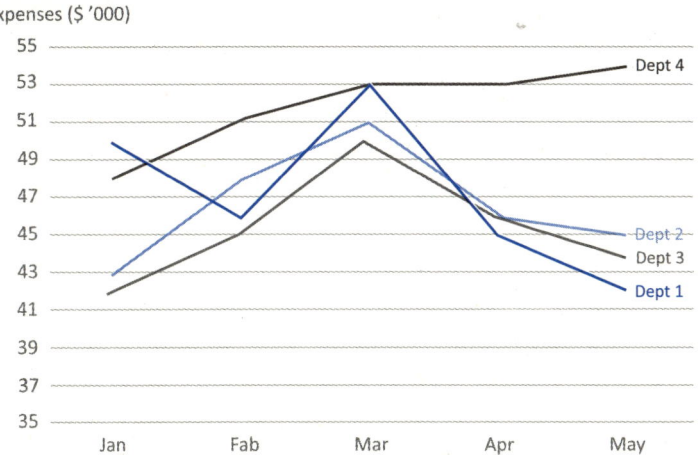

Expenses ($ '000)

Figure 4.13: A line graph instead of a stacked area graph reveals a completely different pattern.

the only category which has a straight baseline, and hence the shape of its trendline actually means something.

The other three categories do not rest on straight lines—the movement of their lines depend not only on their own expenses but also on the movement of the categories below them. For Departments 2 through 4, the change in thickness of their bands as opposed to the movement of their top line is the actual indicator of the trend in their expense. Estimating the change in thickness of each band when the baseline for each is also jagged is visually very challenging. Most viewers tend to look at just the top lines for each category, leading to incorrect interpretations. The rising expenses for Department 4 were completely hidden in the stacked area graph because the peaks in March for the other three departments created a misleading baseline on which Department 4's expenses were plotted.

With these inherent disadvantages, one would expect the stacked area graph to be rarely used, but sadly they can be seen in a wide variety of places. The two real-world examples discussed next epitomize the misuse of stacked area graphs.

The *World Inequality Report 2018,* published by the World Inequality Lab, uses a stacked area graph to show region-wide distributions of income groups (available online[5]). This graph is problematic for multiple reasons. The bottom-most category in the graph (India) has a large section of its population in the lowest-income groups. This creates a huge hump in the graph, and other regions which are stacked on top of India follow the same curve even when their populations are more evenly distributed along the income scale. It is nearly impossible to determine what the actual income distributions are for the other countries when visualized as a stacked area graph.

The second problem with the graph is the unexpected change in scale of the x-axis which denotes income percentiles. Each tick in the scale initially marks a jump of 10 on the percentile scale, but beyond 99 percentile the scale is stretched out. The top 1 percentile occupies the same space as 30 percentile gap in the rest of the graph, leading to misleading trend lines for all the categories!

This double whammy of a graph with its heart-stopping high-intensity colours and multiple design flaws exemplifies what we have been discussing all along—misleading graphs which leave the viewer perplexed and muddled.

While the above graph may not have had an agenda behind the confusing visual, the same cannot be said for the next example. A tech start-up (name withheld) used a graph similar to that shown in Figure 4.14 for a funding pitch. The tagline for the graph stated that all their customer segments were steadily

---

[5] Stacked area graph of population distribution by income, page 53, https://wir2018.wid.world/files/download/wir2018-full-report-english.pdf

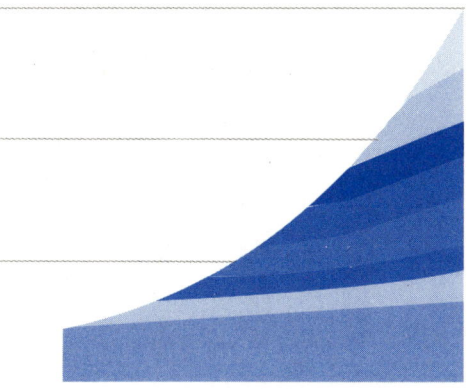

Figure 4.14: Stacked area graph used by a company to claim that all customer segments were growing.

growing. It is clear that the total number of customers of this company have increased, but is it valid to claim that each segment has increased? Probably not. Each line moves upwards, but some categories appear to increase only because they are based on an upward slope. The stacked construct provides an ideal tool for deceiving the audience.

There are only two useful lines in a stacked area graph, the line which represents the bottom category and the top-most line which can be used to interpret—'not' the movement of the top category but the cumulative trend for all the categories put together. However, if the trends for individual categories need to be traced correctly, please do replace stacked area graphs with basic line graphs.

THE EFFECT OF INVERSION

The simplest of changes in a graph can trip up the audience. Sometimes intentionally, but very often unintentionally, a design

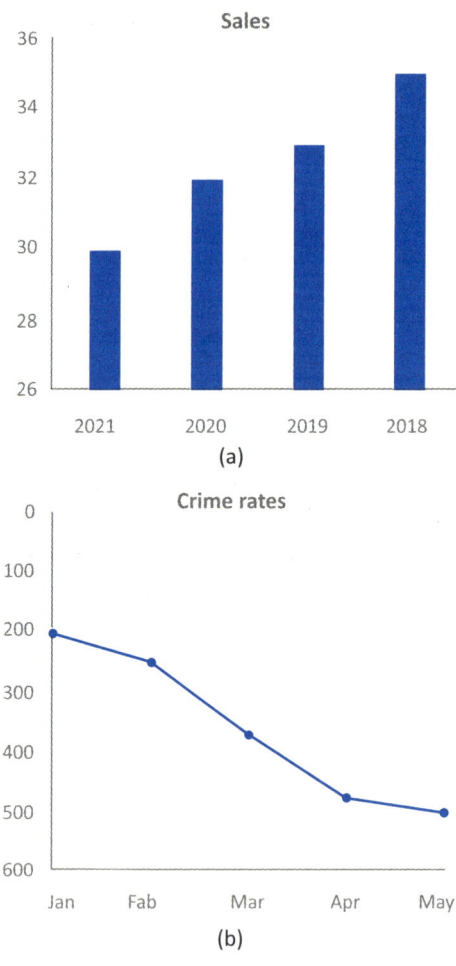

Figure 4.15: Trend graphs on sales over time (a) and crime rates for the past few months (b).

decision which seems innovative or creative can cause the data to be misunderstood. Look at the two graphs in Figure 4.15 on sales trend across years and crime rates for the past six months. Would it be fair to say that sales have increased over the last few years and crime rates have fallen quite dramatically in the last six months?

Look again. Sales have actually 'decreased' over time and crime rates have 'increased' in the last six months! The bar graph on sales has multiple problems. Not only does the y-axis scale not start at zero, time on the x-axis flows from right to left. In the line chart depicting crime rates, the y-axis scale starts from the top instead of the bottom.

In both these charts, the scale of an axis runs opposite to what the viewer expects. In time series graphs, the time scale typically runs from left to right and quantitative scales typically start from the bottom left. While these rules are not set in stone, these conventions are commonly followed and hence are reasonable assumption for the audience to make.

Why, you might ask, should time flow from left to right in graphs? Why not right to left or bottom to top or top to bottom? While a top to bottom flow might still work for a time series, it is not as effective as right to left (see Figure 4.16). This is primarily because most languages are written (and read) from left to right and top to bottom and so that is how we tend to scan a graph as well. If we want to see the data in chronological order, it makes sense to align the data to the natural direction of how we scan information.

Of course, that then begets the question of why we prefer the scale for quantitative axes to start from the bottom left of the graph. A y-axis scale which starts at the bottom aligns with our notion of larger quantitative values being higher than smaller values, while an x-axis quantitative scale which starts on the left is probably preferred because it aligns with our left-to-right scanning bias. Regardless of the reasons, viewers do arrive at graphs with certain predispositions on how the information is going to be arranged, and we need to be cognizant of the same when attempting to present our data in novel ways.

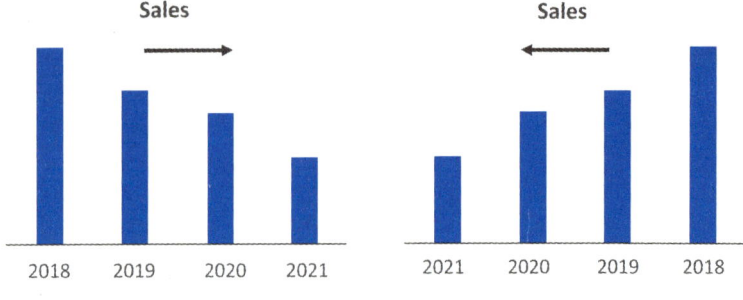

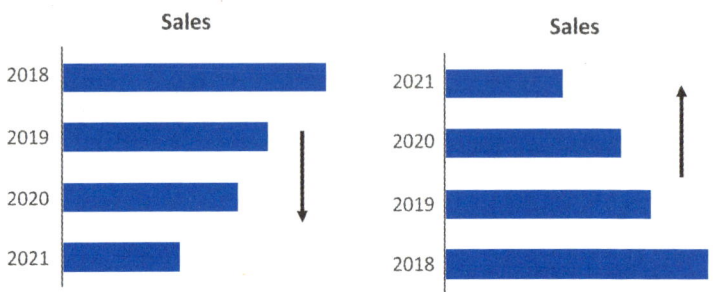

Figure 4.16: Four different arrangements for time series data. The arrow indicates the direction of scan expected from the audience. Left-to-right is the most preferred option and top-to-down the second-most preferred.

This is the double edge of novelty in charts. There should be a very high bar against running counter to convention. Readers do bring their 'baggage' to the chart, and the designer should take that into consideration.

Kaiser Fung, Junk Charts

A few years ago, Reuters put out a graph plotting gun deaths in Florida in the context of Florida's 'stand your ground' law. The graph generated a lot of controversy because it used an inverted y-axis scale for the number of murders committed using firearms. It was criticized harshly. Many readers felt that the graph was misleading because while gun deaths actually spiked sharply after the law was introduced, in the graph, the line representing murders dropped downwards. While the graph is no longer available on the Reuters website, you can see it at this link[6] or google the following: 'Florida gun deaths chart'.

A lot of critique of the visualization targeted the designer of the graph for wilfully misleading the reader, prompting her to put out a clarification. The designer stated that she did not intend to mislead the audience and that she was inspired by another well-known graph on US military deaths in Iraq which used the same reverse axis scale technique. Surprisingly, the Iraq graph had never been called out as misleading.

However, a closer look (refer to the online link given below[7]) reveals certain design decisions which help one graph be effective in spite of the inverted axis, while the other lands up being confusing. The Florida graph has its baseline (x-axis timeline) at the bottom, while the Iraq graph has it at the top. The data in the Iraq graph thus visually hangs from the top baseline.

In both cases, the data is coloured red and the background is white. Since the Iraq graph places many other components in the white space, it is clear that the red part represents the data. This is not very obvious in the Florida graph, leading many viewers to consider the white as the data on a red background, leading to the opposite takeaway. Deliberate or otherwise, the reverse axis scale did cause many viewers to interpret the data incorrectly.

There is however one scenario where a reverse axis scale is actually desirable. Graphs which display ranks should order the

---

[6] https://www.businessinsider.in/politics/this-chart-shows-what-happened-to-gun-deaths-in-florida-after-stand-your-ground-was-enacted/articleshow/30635752.cms

[7] http://www.simonscarr.com/iraqs-bloody-toll

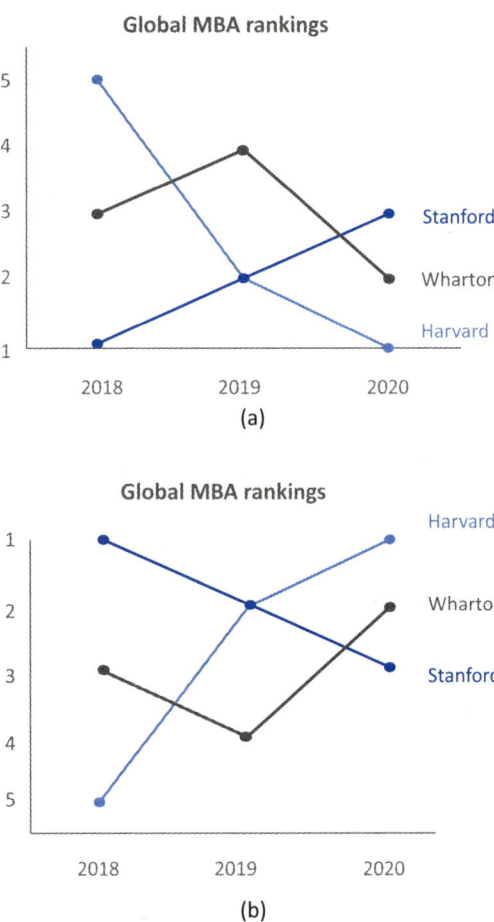

Figure 4.17: A bottom-to-top ranking scale creates confusion (a) while a top-to-bottom scale (b) is more intuitive.

ranks from top to down and not vice versa, which implies that the y-axis scale will actually be in reverse order. As an example, refer to the two graphs in Figure 4.17, which show the rankings of three MBA programmes across three years.

After a quick glance at graph (a), we may conclude that Harvard has dropped in its ranking, while Stanford has steadily improved. However, these interpretations are incorrect because the best rank is placed at the bottom. Graph (b) with its inverted y-axis scale is more intuitive. Using this graph, it is easy to see that the Harvard programme has actually substantially improved in its rankings, while Stanford has dropped from being the best in 2018 to third place in 2020.

We have a preference for ranks arranged from top to down, and visualizations which incorporate this are easier to comprehend.

To sum up this discussion, there is a fine line between deception and confusion, and even the best of intentions may not be enough when conventions are flouted. Play it safe, learn the conventions and design graphs which align with the intuition of the viewer. Remember that your audience deserves a data visualization which is clear and easy to decipher.

DUAL AXIS GRAPHS

Dual axis graphs, also known as superimposed graphs or dual scale graphs, are primary used for depicting time series relationships. There are two scenarios where they are popular: The first involves two categories representing the same unit of measure but which happen to be very different in magnitude. The second scenario involves two unrelated variables with different units of measure.

In the first scenario, since the values for one category are much smaller than the other, plotting both on the same scale makes it difficult to accurately see the trend for either. This is illustrated in Figure 4.18, where domestic and international sales of a company are plotted on the same graph. Since domestic sales are an order of magnitude greater than international sales, using the same scale to fit in both lines makes it difficult to see their

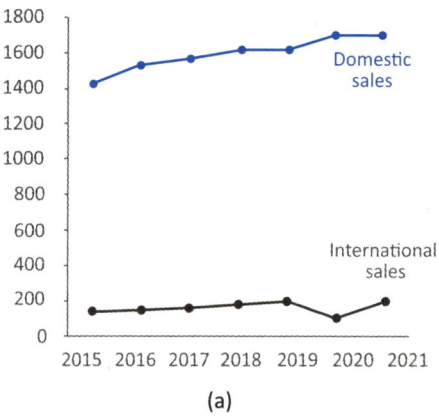

(a)

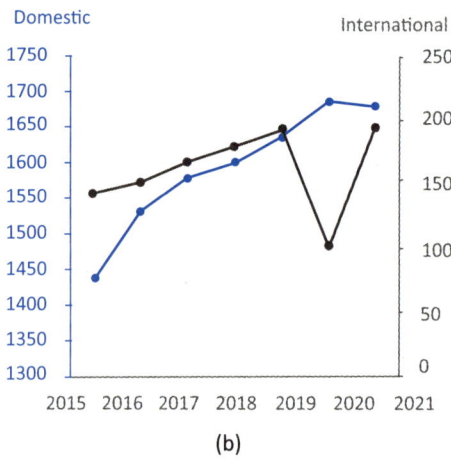

(b)

Figure 4.18: Two variables which defer greatly in magnitude are plotted on the same scale (a) and a dual scale graph (b). The dual scale graph could mislead some viewers.

individual nuances. It may be tempting in such a case to use two different scales (graph (b) in the same figure) so that both scales can be 'opened up' to reveal the trends for each category. With the dual axis graph, we can clearly see the sharp dip in international sales in 2020.

A dual scale graph is essentially the result of two separate line graphs superimposed on each other to create a more compact visual. The danger however with this construct is that the unaware audience might land up visually comparing the magnitudes of the two categories. In our example, the audience may wrongly assume that international sales have exceeded domestic sales for the first five years.

Dual scale graphs are also used to compare the performance of two separate variables over time, for example, the GDP per capita of a country and its happiness score. The aim is usually to understand the relationship (correlation) between the two variables over time. Since these are two unrelated units (USD and happiness score in our example), the dual scale graph is considered by many as a handy construct to compare both variables.

Despite their popularity, the use of dual scale graphs is discouraged by many data visualization experts. In terms of the amount of criticism garnered, they rank lower only to pie charts!

Stephen Few makes a strong case for not using dual scale graphs with bars or dots.[8] Even for an interval scale like time, Few strongly discourages using a dual scale graph because of the muddle they can create in the audience's mind. Haemer's much quoted article, aptly titled 'Double Scales Are Dangerous',[9] notes that these graphs are a 'shortcut to confusion' because they may show things which do not really exist and produce 'a distorted picture if they are not used properly'. Hadley Wickham,

---

[8] Stephen Few, 'Dual-Scaled Axes in Graphs: Are They Ever the Best Solution?' *Perceptual Edge Visual Business Intelligence Newsletter* (March 2008), https://www.perceptualedge.com/articles/visual_business_intelligence/dual-scaled_axes.pdf

[9] Kenneth W. Haemer, 'Double Scales Are Dangerous', *The American Statistician* 2, –no. 3 (1948): 25.

RStudio's chief scientist, has famously stated that ggplot2 does not support dual axis graphs as they are 'relatively hard to interpret' and 'easily manipulated to mislead'.[10] Many others have also advised utmost caution while using dual axis graphs.

So what exactly is the problem with these graphs? Let us dig deeper with a specific example introduced earlier. Figure 4.19a contains a dual axis graph which records the per capita GDP versus the happiness score for India across six years.

The graph suggests that while GDP/capita has steadily increased in India, happiness has actually decreased over time. While this conclusion seems valid, take a look at the other versions of the same graph. While the underlying dataset is exactly the same, the two y-axis scales have been modified, dramatically altering the visual relationship of the two lines. While the lines criss-cross in graph (a), they seem to converge in graph (b) and diverge in graph (c). Moreover, the relative slopes of the lines change across the versions.

Note that these are all valid ways of representing the data which only goes to show that lines crossing each other have no significance whatsoever in dual axis graphs. Both lines in a dual axis graph need to be viewed separately, and comparing the slopes of the lines is meaningless, since either axis can be manipulated to change the slope of one line independent of the other.

Additionally, a naive audience might land up comparing the magnitude of the two lines, which is completely meaningless as well. To add to the above visual red herrings, studies[11] have shown that audiences have trouble decoding which axis

[10] https://stackoverflow.com/questions/3099219/ggplot-with-2-y-axes-on-each-side-and-different-scales

[11] P. Isenberg, Anastasia Bezerianos, Pierre Dragicevic, and Jean-Daniel Fekete, 'A Study on Dual-Scale Data Charts', *IEEE Transactions on Visualization and Computer Graphics* 17, no. 12 (December 2011): 2469–2478.

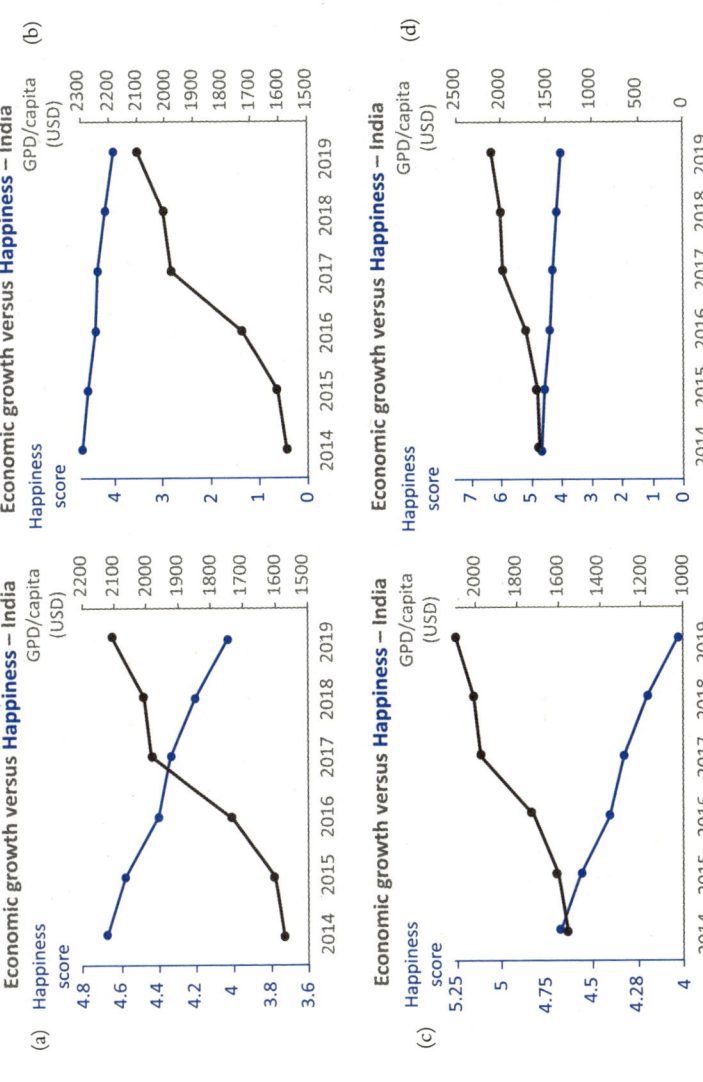

Figure 4.19: Various ways the same dataset can be visualized using dual axis graphs. The individual slopes and criss-crossing of lines can be easily manipulated, leading to different interpretations.

pertains to which line. The last problem can be solved to a large extent by colour-coding the corresponding axis and line, but unfortunately the rest of the problems are enough to trip up most unsuspecting audiences.

What can be done to reduce these drawbacks of the dual axis construct? Since most of the misinterpretations stem from placing both lines in the same graph, one possibility is to use two separate graphs instead of one (Figure 4.20). This mitigates the

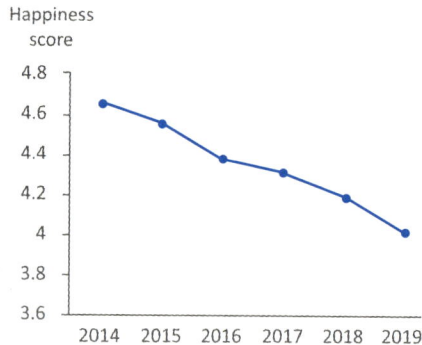

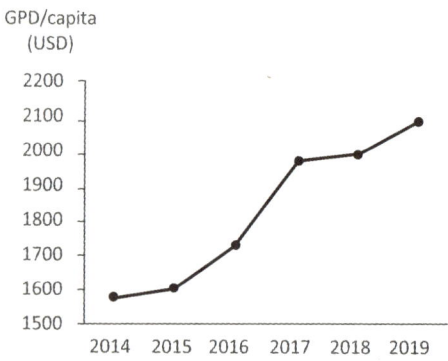

Figure 4.20: Alternative to dual axis graphs—two independent graphs with the same timeline.

possible confusion caused by the false crossover point and also dissuades the audience from comparing the magnitudes of the two unrelated variables.

The slopes of the two lines can still be independently manipulated though. This however can be overcome by another technique—anchoring both variables to a common index so that only one scale is needed. A simple way to achieve this is to show the percentage change for both variables from one particular point in time (see Figure 4.21). While indexing is a commonly accepted solution for comparing the trends of two variables, it does not allow for viewing the actual units involved. Interpreting the trend of variables as an abstract index (percentage change in this case) can be unintuitive for some audiences.

A happy comprise to this situation is to use a carefully designed dual axis graph as depicted in Figure 4.19d. Both scales start from zero and are calibrated such that both lines begin from a common point. This design takes care of the 'false crossover point' problem and 'independently manipulated slopes' problem while also preserving the actual units of the two variables.

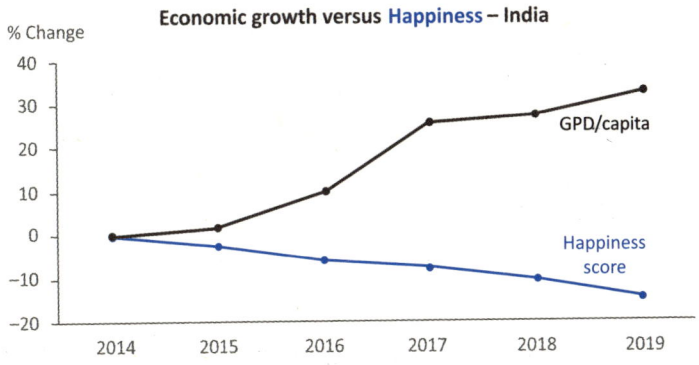

Figure 4.21: Alternative to dual axis graph—indexing the variables to a common scale.

In many ways, this technique is equivalent to using an index; think of it as indexing with benefits.

Another alternative to the dual axis time series graph (and recommended by most visualization experts) is a connected scatterplot. You have already encountered a scatterplot in the first chapter as the ideal construct to study the correlation between two variables. A scatterplot, however, does not capture the time dimension. A connected scatterplot retains the strength of a scatterplot but incorporates the temporal element by joining the dots chronologically.

In Figure 4.22, the scatterplot helps us spot a distinct negative correlation between GDP/capita and happiness in India. The connected scatterplot provides additional information in terms of time—we can infer that while the GDP/capita has steadily increased between 2014 and 2019, the happiness score has correspondingly dipped over the years.

The connected scatterplot has traditionally been used to analyse data, and only more recently has it become popular as a communication tool. Therefore, it is an unfamiliar construct for many audiences and must be designed carefully to avoid confusion or misinterpretation.

In an experimental study on viewers' ability to interpret data presented as a connected scatterplot, Haroz et al.[12] found that the construct has both strengths and weaknesses. The study discovered that while viewers were intrigued by the novel graph, they often assumed that time was flowing from left to right, even if it was actually flowing in the opposite direction. Unless clearly marked with arrows or other methods, the inherent bias of time flowing from left to right took over. They also found that

[12] S. Haroz, R. Kosara, and S. Franconeri, 'The Connected Scatterplot for Presenting Paired Time Series', *IEEE Transactions on Visualization and Computer Graphics* 22, no. 9 (2016): 2174–2186.

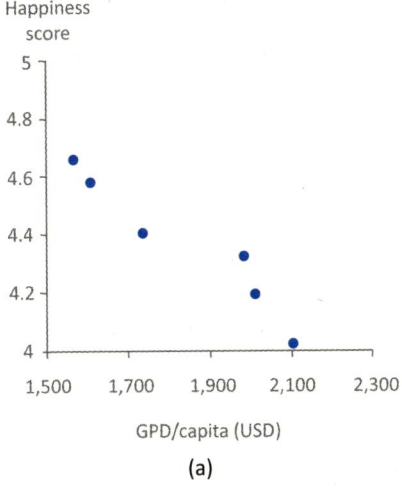

(a)

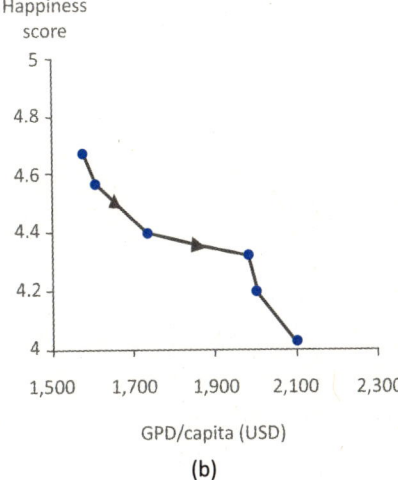

(b)

Figure 4.22: A scatterplot (a) and a connected scatterplot (b) as an alternative to dual axis graphs.

compared to the dual axis graph, audiences were less able to spot correlations in some instances. However, this might have been related to the audience's unfamiliarity with the construct of a basic scatterplot.

Connected scatterplots can be a great tool to bring out a correlation narrative which plays out over time. If you decide to use one, make sure that (a) your audience is familiar with a basic scatterplot construct and (b) the direction of time's flow is clearly marked on the chart.

In summary, dual axis graphs for paired time series data can be effective if carefully designed. One technique is to start both y-axes scales at zero and adjust the scales such that both lines have a common starting point. Some alternative constructs we discussed are recapped below:

- Index the variables to a single scale

- Separate the lines out into individual graphs with the same time scale, sometimes also called faceting (place the graphs one below the other for easier comparison)

- Use a connected scatterplot instead

IN CONCLUSION

This chapter covered a range of constructs and design choices to avoid spurious interpretations of trends. You now understand the importance of consistent scales and the significance of understanding the conventions your audience might be used to. The knowledge of the limitations of stacked area charts and dual axis graphs along with alternate constructs to use instead of these will help you create honest graphs which bring out the true patterns in the data.

The next chapter switches gear to delve into finer graph design principles which help in building clear, intuitive constructs, which in turn makes the data easier to interpret.

CHAPTER 5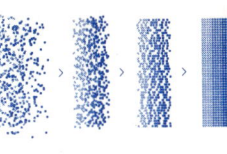

DESIGN CHOICES FOR ACCURATE DATA INTERPRETATION

What is an effective graph? As Naomi Robbins puts it, *'Effective graphs improve understanding of data. They do not confuse or mislead.'*[1]

Our discussion so far has led us through many different visualization constructs and design choices to create an effective graph. In this chapter, we dive deeper into the nuances of effective design. The guiding principle in this chapter is to make it easy and intuitive for the audience to accurately interpret the data. Any element which confuses your audience or distracts them from the key insights should be identified and corrected.

We will first look at some generic principles applicable to all graphs and then concentrate on specific graph types, including pies, bars and lines.

---

[1] Naomi B. Robbins, *Creating More Effective Graphs* (Hoboken: Wiley InterScience, 2005).

GENERAL DESIGN GUIDELINES FOR ALL GRAPHS

Tufte's principle—'Above all else, show the data'—is a handy guideline to make your graph clear and easy to read. Always make sure that there is one and only one protagonist in your visualization—the data. Other elements such as the axes, title, legend, axes titles, grid lines and annotations are called supporting components for a good reason. They are there to support the actual data and should be given less visual emphasis compared to the data.

A quick example will help illustrate this guideline. The graph in Figure 5.1 shows two versions of the same construct—a line graph denoting the medal tally for four schools across eight years.

Schools 1 and 2 had minor fluctuations in their medal tally. School 3 has improved consistently, while School 4's tally has dropped drastically over the years.

Notice how the second version is not only more pleasant to look at but also makes it easier to quickly grasp the trend in medals won over the years for each school. The same takeaways can be gleaned from the first version as well but perhaps at the cost of a headache or at least some mild disorientation.

Take a couple of minutes to note down all the design differences that make the second version more effective.

Did you notice the many distractions in the first graph which draw attention away from the data, which in this case are the four lines? These design elements are sometimes obvious but often subtle and deserve a comprehensive discussion.

Shaded background: To keep the attention on your data, the background of the graph should ideally be transparent or a very light colour. A dark background is not only distracting, but it can also actually come in the way of interpreting the data.

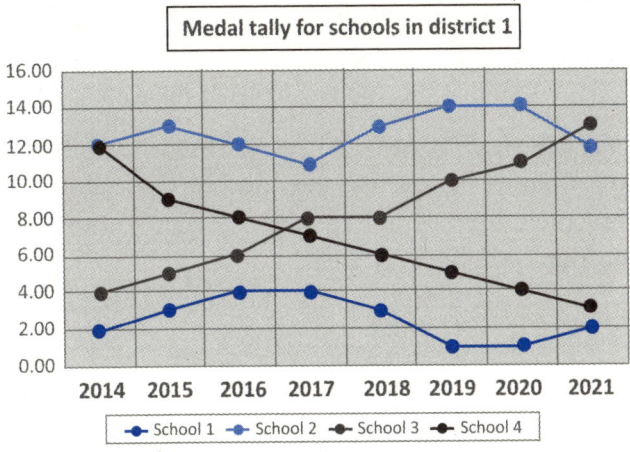

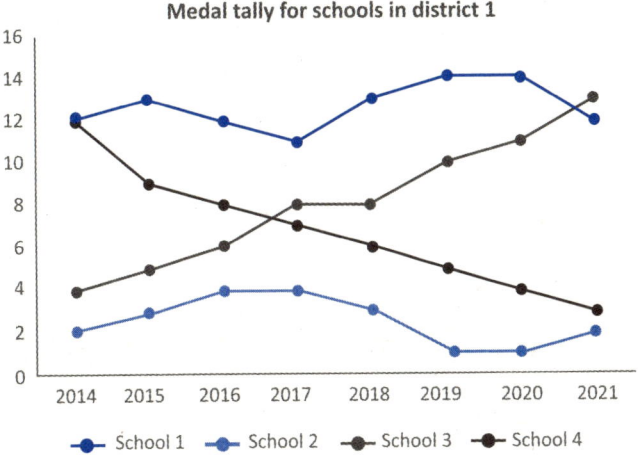

Figure 5.1: Cluttered (top) and non-cluttered (bottom) versions of the same line graph.

Grid lines: Grid lines originate from the era when charts were drawn by hand on graph paper. They are meant for increasing the precision of data interpretation and are an overkill for most graphs today which are machine-drawn. Grid lines are useful in a small set of scenarios, for example, when we want to compare bars of similar height or accurately judge values for data points which are far away from the axis in a wide graph. If you decide to use grid lines, make sure that they are subtle. Choose thin lines in a shade of light grey, so that they do not obstruct the data.

Boxed title and legend: Drawing a border around any element is an established way to call attention to that element. Borders around titles, legends or annotations emphasize that particular supporting component at the cost of the emphasis placed on the data. Avoid all borders including a border around the graph itself or choose a thin grey style which does not stand out.

Precision of numbers: Unnecessary precision in numbers (scale, legends, etc.) not only makes them more difficult to interpret but also takes up precious space which can be allocated to the actual data. In our example, it is completely meaningless to show medal counts up to two decimal places! Precision up to two decimal places is the default number format in some graphing software, so watch out for this particular blooper.

Font considerations: Bold colours and/or a large font size for the text and numbers can add an extra layer of distraction. Make sure you use a medium grey font colour for the axis scales and a size that is just big enough to decipher. Additionally, using the same font type for all elements ensures anonymity for your supporting components.

These are some generic design principles to keep in mind while creating an effective visualization. Our discussion now moves to specific design considerations for the three most popular constructs: pie charts, bar graphs and lines graphs.

GETTING THE PIE TO WORK FOR YOU

As noted in the introductory chapter, many data visualization experts harbour a strong dislike for pie charts. Studies have also established that they are ineffective for many common scenarios.

However, despite the heavy criticism pelted down on pie charts, why do they continue to be so popular? The pie chart does have a hidden strength—it is a very intuitive construct for displaying part-to-whole relationships.

A part-to-whole relationship can typically be worded as a fraction or percentage; for example, 85 per cent of the customers in our survey were happy with the new after-sales programme. For visualizing a part-to-whole relationship, using a single shape as 100 per cent and demarcating it's parts into percentages works well at an intuitive level. There is a good reason why many math curriculums use a pizza cut into slices to introduce children to the concept of fractions.

While bar and line charts do not fit the bill of a whole and its comprising parts, pie charts and 100 per cent stacked bar charts fulfil the requirement. Figure 5.2 depicts the part-to-whole example discussed earlier using three different constructs: bars, 100 per cent stacked bar and pie chart. When it comes to capturing the essence of a part-to-whole relationship, both the 100 per cent stacked bar and the pie have a definite advantage over individual bars.

Hundred per cent stacked bar charts employ 1D coding, which means that the audience is required to compare the lengths of the various segments which we already know is easier than comparing areas. For pie charts, however, the jury is still out on whether the viewer compares the areas, angles or arc lengths of different slices. Studies have conjectured that it is probably a combination of all three elements but have on the whole been inconclusive.[2]

---

[2] Drew Skau and Robert Kosara, 'Arcs, Angles, or Areas: Individual Data Encodings in Pie and Donut Charts', *Computer Graphics Forum* 35, no. 3J (2016): 121–130.

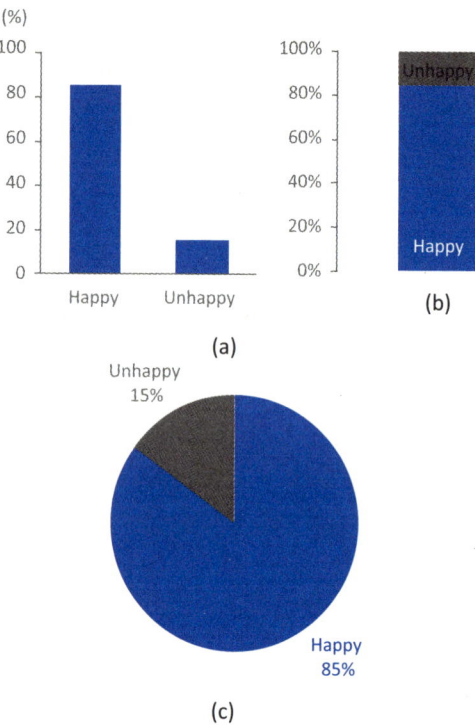

Figure 5.2: Part-to-whole relationship depicted using bars (a), 100 per cent stacked bar (b) and pie chart (c).

However, one definite advantage of the pie chart is our clear understanding of the halfway point, which is due to our familiarity with the clock construct and the straight vertical line depicting 6 o'clock. Not only the halfway point, a quarter (3 o'clock) and three quarters (9 o'clock) are also easily distinguishable in a pie chart. This is not the case for other constructs like stacked bars, where our estimates of 25 per cent, 50 per cent and 75 per cent are not as accurate. To leverage this advantage though, the first slice of the pie chart should ideally start at the 12 o'clock position.

Figure 5.3 depicts two pie charts—the fraction of the blue slice is more difficult to estimate in the first one. In the second pie

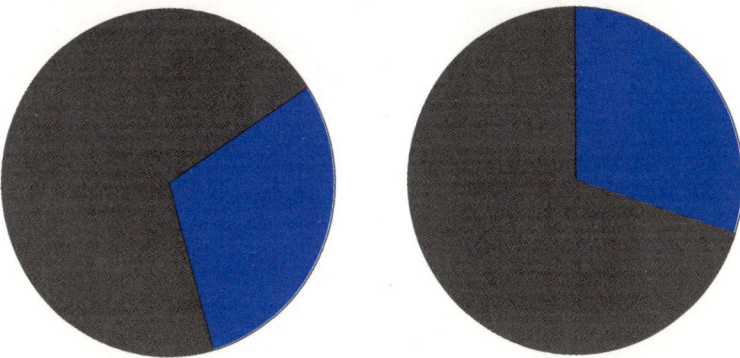

Figure 5.3: Starting the first slice at the 12 o'clock position (right) makes it easier to estimate the fraction of the slice.

chart, one side of the blue slice is aligned with the 12 o'clock position, making it easier to decipher that it is a little more than one fourth.

Figure 5.4 illustrates the distinct advantage of a pie chart in comparison to a stacked bar chart. Using the pie chart in the top row, it is easy to estimate that the blue category is a little more than half. The same conclusion is not that obvious in the corresponding stacked bar. Again, in the bottom row, the pie chart makes it effortless to estimate that the blue category is a little less than one fourth. The stacked bar, however, does not provide the same affordance.

While pie charts are powerful constructs for depicting a part-to-whole scenario with binary categories, we have already noted that their effectiveness diminishes rapidly if there are more than three or four categories.

There are, however, certain scenarios with multiple categories where pie charts are the right choice. A 1991 study[3] at the

---

[3] I. Spence and S. Lewandowsky, 'Displaying Proportions and Percentages', *Applied Cognitive Psychology* 5, no. 1 (1991): 61–77.

Figure 5.4: It is easier to estimate a fraction in a pie chart versus a 100 per cent stacked bar.

University of Toronto found that pie charts work better than bar charts when the sums of two groups of categories have to be compared. Notice in the example in Figure 5.5 how the pie chart makes it easier than the corresponding bar chart to conclude that (A + B + C) < (D + E + F). If you do use a pie chart for a similar situation, ensure that the slices which need to be added up are placed next to each other. Additionally, choosing the same hue for slices which need to be added aids in the cognitive task of seeing them as one group.

The above example is actually just a variation of the binary categories scenario, as ultimately only two resultant categories need to be compared. For most other situations with multiple

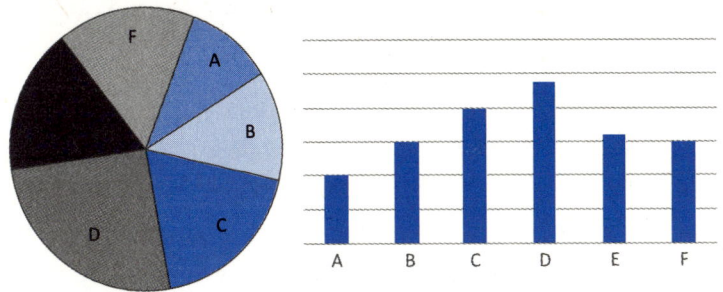

Figure 5.5: This pie chart is more effective than the corresponding bar chart to convey that (A + B + C) < (D + E + F).

categories, remember that there is strong evidence in favour of bar charts as compared to pie charts.

THE HONEST BAR GRAPH

You already know by now that bars which do not start at a zero baseline are grossly misleading. While this is the most obvious and common blunder, there are other design aspects to be cognizant of while striving for a truthful bar chart. A few important considerations are the use of data labels, relative positioning of bars and colour choices.

Data Labels: To Use or Not to Use

The general consensus among visualization experts is that graphs should be used to bring out patterns in data, and tables are the correct choice if individual values need to be looked up. This school of thought also implies that since data labels are a means to find out the exact value of a particular data point, they are superfluous for graphs. In the event that exact values are also needed, you can use a graph 'without' data labels but supplement it with a separate table containing the values. In this way, the labels do not obstruct from deciphering the pattern in the data while, at the same time, the exact values are easily accessible in the same view.

While this argument holds for many scenarios, there are a few exceptions to be aware of. Certain industries(e.g., the consulting domain) are used to mentioning the values of all data points in a bar graph. In fact, the practice is so common in this industry that visualization enthusiasts have given these graphs a special name—grables (graph + table). If your audience is used to a particular way of deciphering data, it might make sense to not rock the boat too much. With careful attention to the position of the labels, a grable can still be used to accurately visualize the data.

As shown in Figure 5.6, positioning the labels outside the bar can interfere with accurately comparing the lengths of the bars. A more effective practice is to place the labels inside the bars so that the perceived heights of the bars are not skewed. Also, to ensure maximum visibility, the text colour of the labels should be in contrast to the bar colour. Moreover, if all the bars are labelled, then the quantitative axis is redundant and can be omitted.

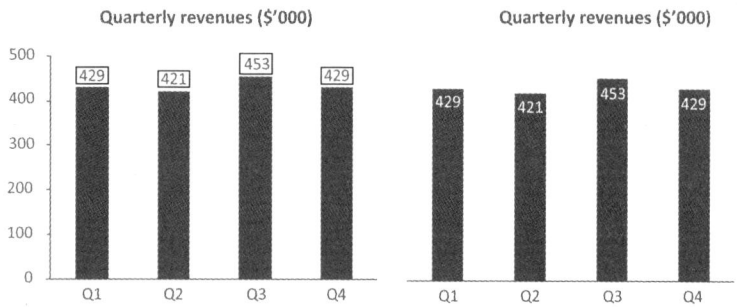

Figure 5.6: Data labels placed outside the bar (left) can hamper the judgement of bar lengths. The correct place for data labels is inside the bar (right).

Instead of labelling all data points, selective labelling can be used as an effective strategy for drawing attention to a particular part of the graph. When a specific data point needs to be called out as part of the message or small differences in values between two or more data points need to be emphasized, selective labelling can come in handy.

While on the subject of data labels, labelling all the data points in a dense line graph should be avoided. The labels add clutter and can make it more difficult to see the trend in the data. A separate supplementary table as shown in Figure 5.7 would be the ideal choice for situations where the exact values are also needed.

Positioning of Bars

Bar charts are the default construct when the primary aim is to compare values for a few different categories, also called a nominal comparison. A nominal comparison does not entail that the categories be listed in any particular order. For example, if a company wants to compare its profits in the current quarter to its main competitors, the bar graph in Figure 5.8 might be considered suitable enough for the purpose.

However, the above graph is less than ideal because of a phenomenon called the Ebbinghaus illusion. The Ebbinghaus illusion[4] is an optical illusion where the perception of the size of an object depends on surrounding objects. The best known example of this illusion is illustrated in Figure 5.9, where both the blue circles are exactly the same size but one appears bigger than the other because of the surrounding circles.

---

[4] B. Roberts, M. G. Harris, and T. A. Yates, 'The Roles of Inducer Size and Distance in the Ebbinghaus Illusion (Titchener Circles)', *Perception* 34, no. 7 (2005): 847–856.

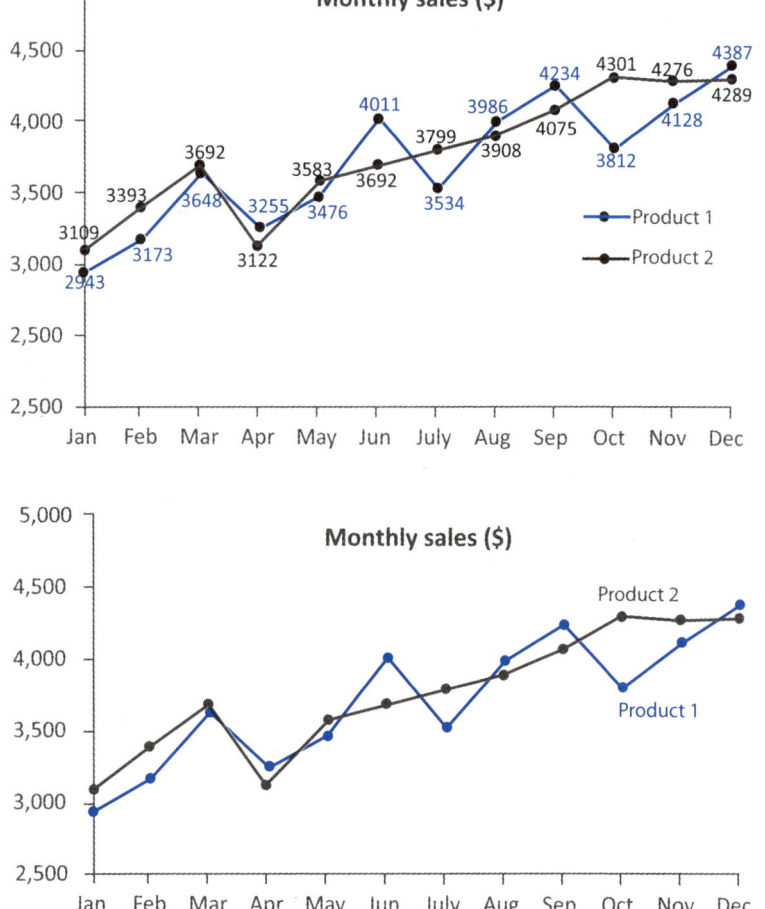

Figure 5.7: Data labels on the lines (top) make the graph too cluttered. A stand-alone table accompanying the graph(bottom) achieves both objectives of showing the trend and access to exact values if needed.

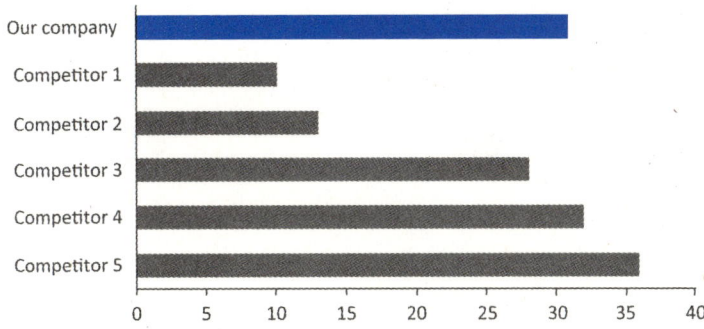

Profits this quarter ($ '000)

Figure 5.8: A bar graph which compares performance of companies. Can this be improved upon?

Positioning a long bar next to a small one can create a similar illusion of it appearing longer than it is. While the effect of the Ebbinghaus illusion has not been systematically studied in bar graphs, we can now see why the bar graph in Figure 5.8 could create an illusion of the blue bar being longer than it actually is.

In any case, placing the blue bar closer to bars of similar length will aid in more accurate comparisons. Sorting the companies according to performance is a simple solution to make it easier to compare the data points. Figure 5.10, with the companies sorted in descending order of profits, makes it clear that 'our company' had slightly lower profits than the two top competitors.

Choosing Colours Judiciously

Choosing the appropriate set of colours for various components of a graph is critical for its effectiveness. On the subject of colouring your graph, Tufte says, 'Above all, do no harm.'

Using lots of colours in your visual can make it attractive and visually captivating but can also confuse the viewer, as they

Figure 5.9: An example of the Ebbinghaus illusion: The blue circles are identical but appear of different sizes because of the surrounding circles.

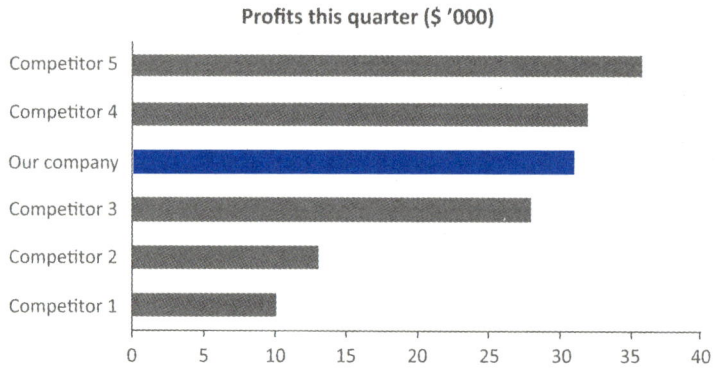

Figure 5.10: Sorting in ascending or descending order helps in accurately comparing bars.

may not know which part to focus on. An effective strategy is to start with a colourless graph and colour only specific parts with the intention of bringing out the desired message in the data.

There are many good resources freely available to help you choose the right colour palette for a particular graph. Color

Brewer[5] is a particularly good source for choosing colours for maps and graphs.

Many of my students have an engineering background with limited exposure to design elements. I often get asked if they can just use the default colouring scheme of a graphing software like MS Excel. I get where the question is coming from. My purely technical education ensured that for many years I only created colourless charts for my articles in scientific journals which printed in black and white. 'Charts' and 'colours' weren't two words I would have uttered in the same sentence. The study of data visualization as a field subsequently opened up the world of colours for me and, with it, the power of strategic colour choices.

So my answer to the students' question is always a resounding 'No'. Unfortunately, many of these software tools have poor colour pallets as their defaults and, moreover, the tools themselves do not understand the purpose of your visualization, only you do! Using colour judiciously, to bring out the intended message in your graph, is a skill worth honing.

For a simple bar graph, all the bars should ideally be of the same colour, as this aids in the most accurate comparison. The colour of a particular bar should only be changed if there is a special reason to call attention to it. In the previous example, the bar representing 'our company' was a different hue than the others, since it needed to be emphasized.

In a clustered bar graph, it is common practice to use a unique colour to represent each category. If equal emphasis needs to be given to each category, it is important to select colours of similar intensity. In the first graph in Figure 5.11, both colours have about the same intensity, leading to equal prominence for national and international sales. In the second graph, however,

---

[5] https://colorbrewer2.org

Figure 5.11: Colours of the same intensity (top) should be used for equal emphasis of categories. A more intense grey (bottom) inadvertently emphasizes international sales.

international sales unintentionally stand out more because of the difference in the intensity of the two shades.

The converse is also true. Increasing the intensity of a colour is a simple and non-messy way to draw attention to a particular bar or set of bars. This technique is particularly helpful when the colour of the said bar cannot be changed because it denotes a particular category. In Figure 5.12, the exceptionally low international sales for Product 3 are called to attention by using both a subtitle for the graph and a darker shade of grey for that particular bar.

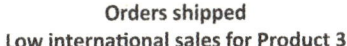

Orders shipped
Low international sales for Product 3

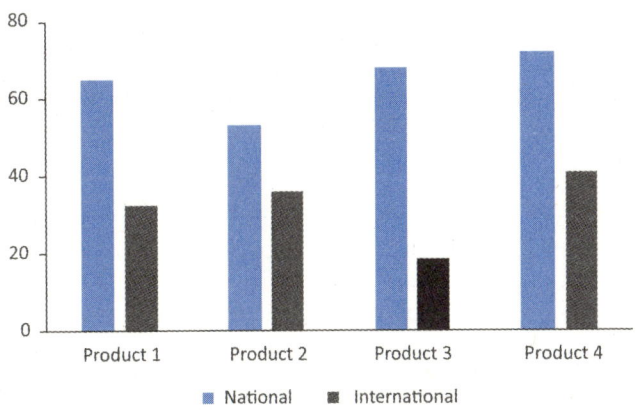

Figure 5.12: Use a shade with higher intensity to make a particular bar stand out.

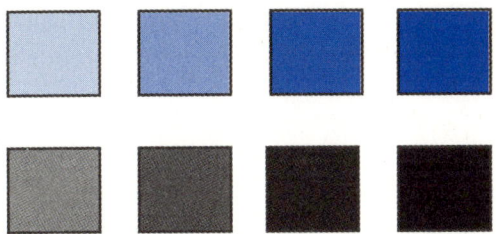

Figure 5.13: Maintain colour palettes with matching intensities for different categories and use a higher intensity to highlight a particular data point.

While deciding how much emphasis different parts of your graph should get, it is handy to maintain colour palettes of matching intensity (as shown in Figure 5.13) to choose from. This will help ensure that categories requiring the same amount of emphasis are assigned colours of the same intensity. Then if a particular bar needs more emphasis, the shade with a higher intensity of the same colour can be chosen from the palette.

In addition to the colour of the bars, careful attention should be given to the background as well. A uniform light-coloured background or no colour at all is guaranteed to work well for most scenarios. A light background ensures that the most important component of the graph, which is the data, stands out well.

Avoid any kind of shading, patterns or objects in the background. These are superfluous and can distract the viewer from the data. Shaded backgrounds can actually come in the way of interpreting the data correctly. The shaded background in the bar graph in Figure 5.14 unintentionally emphasizes the data for Region 1, while the data for Region 5 is almost invisible.

Another common yet inadvisable practice is to use fill patterns for bars. In the event that solid colours cannot be used, resist the temptation to use fill patterns. It is far better to use one hue of different intensities instead. Fill patterns similar to the alternate black and white lines shown in Figure 5.15 are known to create what are called moiré vibrations. Moiré vibrations can make some viewers feel sick or dizzy and in extreme cases are known to cause seizures.

While on the topic of colour choices, choropleth maps are worth a short discussion. Choropleth maps usually use colour gradients to display quantitative values by region. These gradients

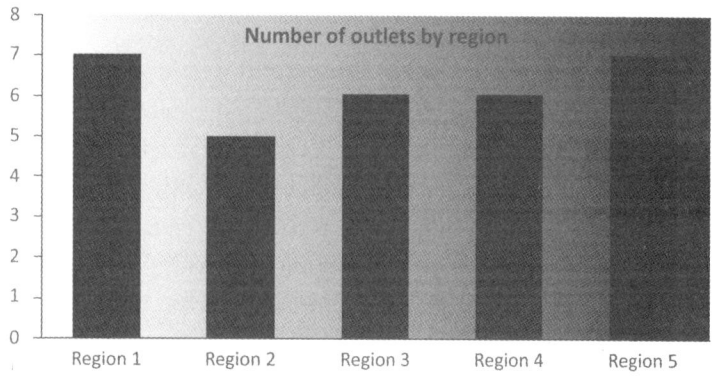

Figure 5.14: The shaded background inadvertently highlights Region 1 and hides Region 5.

Figure 5.15: Avoid fill patterns (top graph) as they can cause some people to feel uneasy or dizzy. Use solid colours (bottom graph) instead.

could represent sequential scales (sale amounts per region, for example) or diverging scales (profits and losses per region). Sequential scales typically work well with a single hue, while diverging scales are best represented by two contrasting hues which increase in intensity in both directions. Figure 5.16 illustrates both these possibilities.

When choosing gradients which represent quantitative values, care should be taken to start the sequential scale with the lightest

Sequential scale

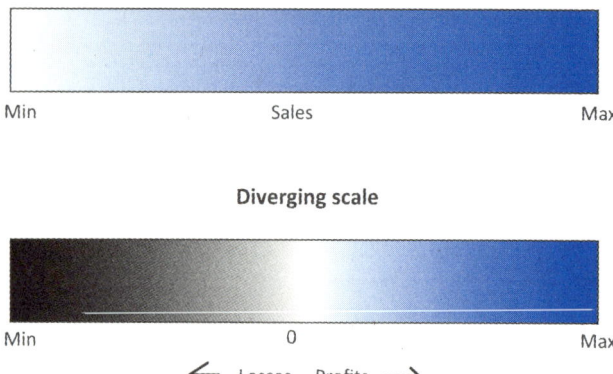

Min   Sales   Max

Diverging scale

Min   0   Max

← Losses   Profits ⟶

Figure 5.16: A sequential scale should use a single hue, whereas a diverging scale works well with two contrasting hues.

shade and move progressively to darker shades. Similarly, for a diverging scale, the midpoint or neutral point should be in the lightest shade and each colour's intensity should increase on both sides. Violating this simple principle can lead to a very confusing choropleth map.

Figure 5.17 contains a choropleth map where 'lighter' shades represent 'more' sales revenue. This colour scheme runs counter-intuitive to what most audiences will assume when they first see the map. They will then have to repeatedly invert their default assumptions of the performance of each district in the map, leaving limited cognitive capacity for the actual analysis.

As noted earlier, your visualization will genuinely aid your audience in understanding the data quickly and correctly if the implicit assumptions and intuitions of the audience are incorporated into the design of the visualization. Creating a visual which runs counter to intuition is much like trying to row against a strong current—a lot of effort for perhaps no gains. Such endeavours are best left to those in the mood for a bumpy adventure!

Sales ($ '000)

	0–10
	10–20
	20–30
	30–40

Figure 5.17: Most audiences assume darker colours as representing larger quantitative values. The colour scale above reverses this convention, creating a misleading graph.

THE TRUTHFUL LINE GRAPH

Line graphs are the default choice when trends over time need to be visualized. Line graphs literally 'join the dots' so that the overall picture of the fluctuations of a particular variable becomes clear. Three important aspects for an effective and truthful line graph are discussed further—the line smoothing feature, legends and the design and use of markers.

Smoothened Lines: Too Much of a Good Thing?

Joining the dots with straight lines can often lead to jagged edges and sharp corners in a line graph. The line smoothening feature available in many visualization software tools smoothens out the edges and creates a more aesthetic graph. The aspect of continuous change to a variable is also better captured with a smoothened line. Figure 5.18 shows both the straight line and smoothened line versions of a graph which plots monthly profits of a company.

While the visual appeal of a smoothened line is difficult to refute, they may in some cases compromise the integrity of the data.

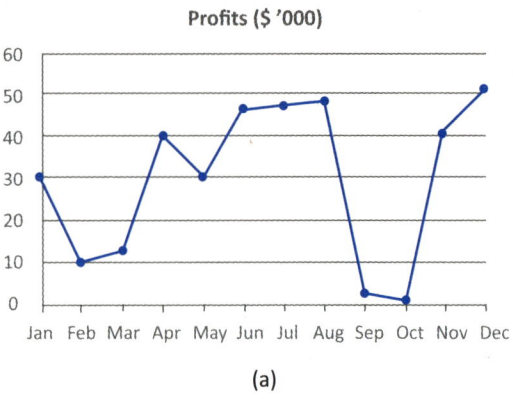

(a)

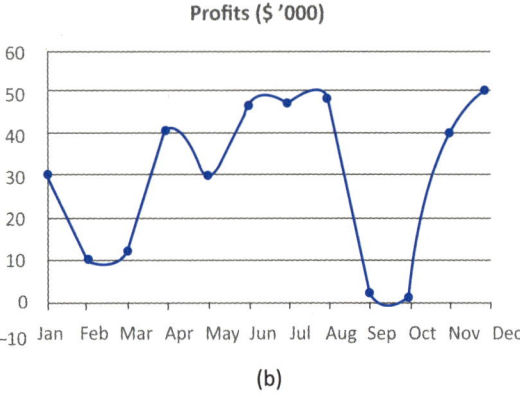

(b)

Figure 5.18: The line smoothening feature can sometimes concoct data, like the losses depicted between September and October in Graph (b).

Consider what happens in the example in Figure 5.18. The straight line graph tells us that profits rose steadily from June to August, while in the smoothened line graph, the profits appear bumpy, which is not true. Additionally, the smoothened line curves upwards between July and August giving, the impression that profits touched 50,000 which is clearly not the case. An even bigger problem is when the profits line inadvertently crosses zero between September and October. Profits dipped dangerously low in September and October, only to recover the following month. The smoothened line however misleadingly conveys that the company slipped into losses in the fall quarter. In a bid to smoothen the lines, data has been concocted out of thin air!

Lines graphs are more accurate when straight lines are used to connect the data dots. Unless there is a compelling need to use the smoothening feature, straight lines and sharp edges are less likely to cause the underlying data to be misinterpreted.

Legends for Line Graphs

A line graph which represents data for multiple categories could typically use a unique colour or marker shape for each category. Such a graph would then need some way to help the user identify each colour or shape with the corresponding category name. The most common solution and one that is used by many software tools is to create a legend (also called a key) to list down the categories and the corresponding colours/shapes.

Figure 5.19 depicts four different versions of a line chart which plots ice cream flavour preferences over the last few decades.

From the graph, we can decipher that while vanilla has been a popular flavour through the ages, chocolate has steadily gained popularity and overtaken vanilla as the most popular flavour. Strawberry, on the other hand, was initially as popular as chocolate but has become less popular in the last twenty years or so.

In which version was it easiest for you to identify the individual flavours and then interpret the data? Most users prefer the last version (version 4) where the lines are directly labelled. Omitting the legend and directly labelling the lines vastly increases the readability of the graph. Also, a subtle but effective design choice is to use the line's colour for the category name. This ensures that criss-crossing lines do not create confusion about which name belongs to which line.

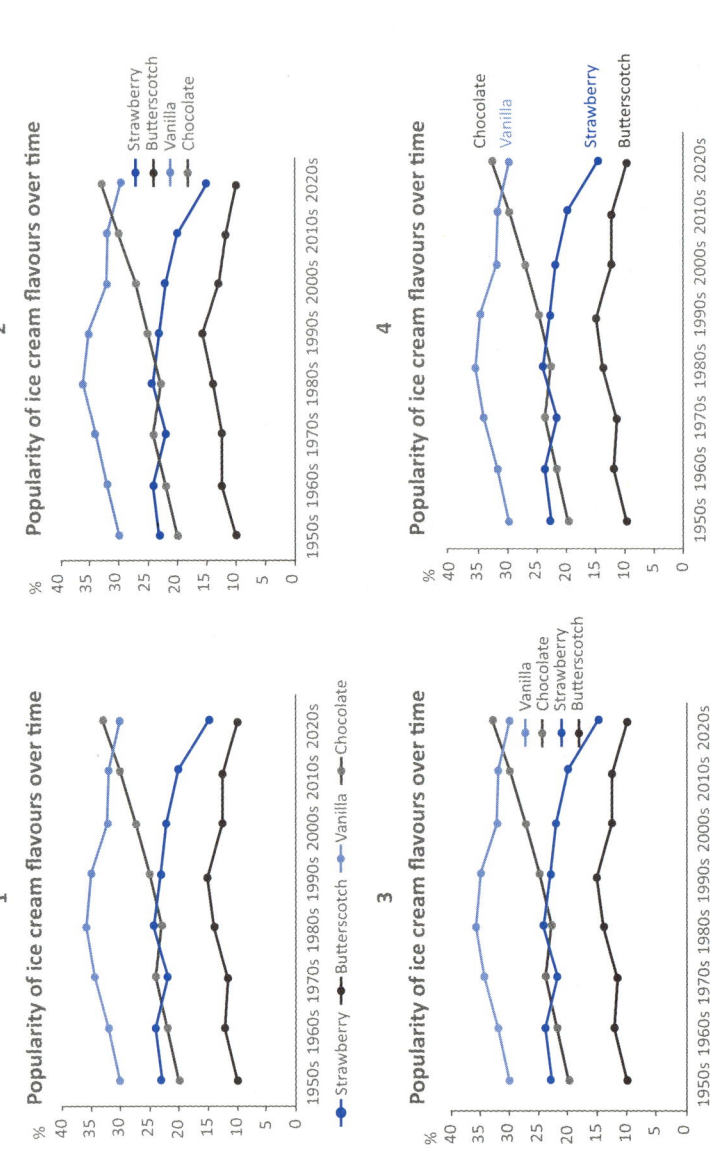

Figure 5.19: Which legend design makes it easiest to understand the ice cream flavour represented by each line?

Data source: The secret worldwide group of ice cream connoisseurs.

Why is labelling the lines directly so much more effective? When data needs to be processed, our brain stores the required data in what is called the 'working memory'.[6] This is much like the random access memory (RAM) of a computer, where the data that needs to be computed is stored. However, unlike the RAM of a machine which can store relatively vast amounts of data, the human working memory can only store up to four chunks of information at a time. What this means for someone trying to interpret the first three versions of the line chart in the above example is that the working memory has to switch between the different chunks of information needed to interpret the graph— the four lines (as four individual chunks of data) and the legend elements (four more chunks). In the fourth version where lines are directly labelled, it is easier to see the line and corresponding category name in one view, as one chunk of data. This means that directly labelling the lines makes it easier for our working memory to store all relevant information simultaneously, leading to more efficient processing of data.

What if for some reason the lines cannot be directly labelled? Sometimes with a larger number of lines, there just might not be enough space on the graph for legible category names, especially if the category names are long. In that case, a legend can be used but ensure that to the extent possible, the order of the names correspond to the order of the lines. Version 3 implements this, which vastly improves the readers' ability to relate the names to the corresponding lines, compared to versions 1 and 2. Version 1 is the default legend design which was created by the software, and it almost feels like a puzzle the user has to solve before they can start interpreting the data. Version 2 is

---

[6] Jeanne Ellis Ormrod, 'Key Components of the Human Memory System: An Overly Simplistic yet Useful Model', in *How We Think and Learn* (Cambridge University Press, March 2017).

marginally better than version 1 because at least the orientation (vertical) matches the data. However, it is not as helpful as Versions 3 and 4 in enabling the viewer to quickly identify the names of the four lines.

Markers in Line Graphs

An important design decision in the construction of line graphs is whether to include markers in the line. The marker is nothing but a shape which denotes an explicit data point. Note that the purpose of a line graph is to follow the trend of a particular variable. Ask yourself if individual markers are really needed to point out the exact locations of the underlying data. More often than not, markers are superfluous and only add clutter to the graph.

The line graph in Figure 5.20 denotes primary energy consumption patterns for 4 selected countries over a span of almost 55 years.

Notice how the markers make the graph look clumsy and distract from following the actual trend of the lines. Does the audience need to know the position of each data point in this graph? Probably not! The same line graph but without markers (Figure 5.21) is much more pleasing to look at, and the trends are easier to comprehend.

There are, however, certain cases where markers can come in handy. If the medium does not allow colours, for example, markers can instead be used to distinguish between categories. Markers can also be used if there are relatively fewer data points in each line and the context demands that the exact location of each reading be clearly distinguished. Markers can also be selectively used if a particular data point needs to be highlighted as part of the narrative of the visualization.

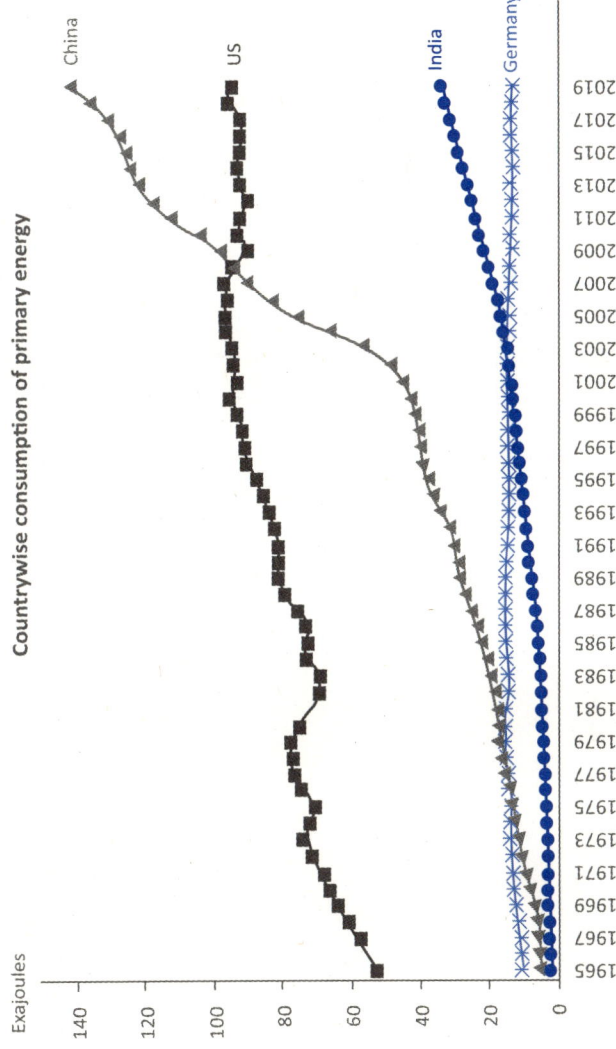

Figure 5.20: Unnecessary markers add clutter and distract the viewer.

Data source: https://www.bp.com/en/global/corporate/energy-economics/statistical-review-of-world-energy.html

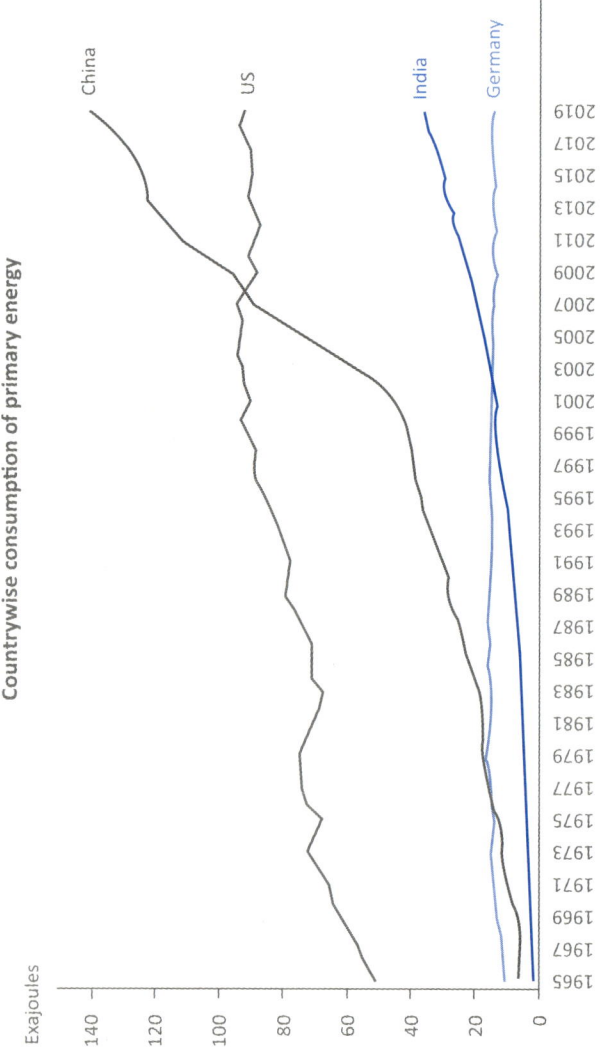

Figure 5.21: Removing the markers makes the data easier to look at and interpret.

The topics covered in this chapter should help the reader appreciate the power of a well-designed graph and the nuances of various design choices and their ramifications.

The next and final chapter of this book deals with visual perception. Understanding how the human visual system works can help draw the audience's attention to certain parts of visualization. This knowledge can help you guide the audience to see a specific message in the data, making it easier to convey the intended data narrative.

CHAPTER 6

TELL YOUR DATA STORY

As goes the popular quote: a picture is worth a thousand words—and a well-designed graph is that picture which can drive forward your data narrative. However, when data insights are communicated via graphs, the interpretations or takeaways from the graph can be as varied as the number of viewers. How do you ensure then that your audience does not miss 'that' particular insight you wanted them to notice?

For an example, look at the line chart in Figure 6.1. This chart contains sales trends over the last six years for an e-commerce

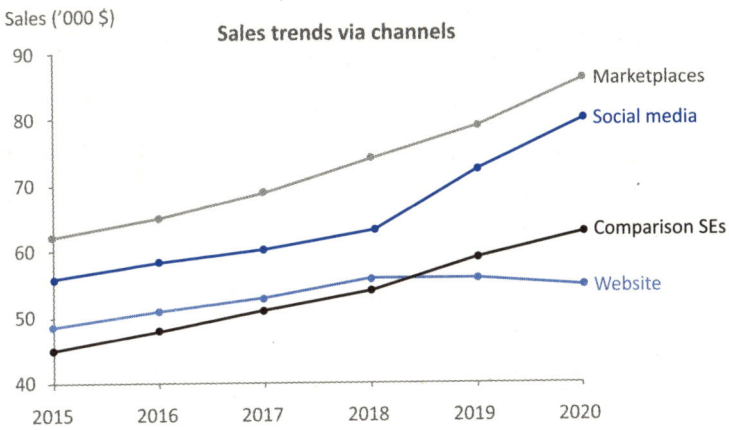

Figure 6.1: A graph can be interpreted in many different ways. How many insights can you generate from these trend lines?

start-up. The company primarily operates through four digital channels: its own website, marketplaces like Amazon, social media and comparison search engines.

What are the different takeaways you can observe in this visualization? One observation could be that overall sales figures are on an upward trajectory. Another observation could be that marketplaces generate the most sales. Yet another takeaway could be that revenue via social media is increasing faster than other channels.

Now suppose the actual purpose of the graph was to show that while other channels have been doing well, website sales have actually declined over the last couple of years. Perhaps this is alarming and demands a relook at the overall marketing strategy for the company's web portal. Take a moment to think about possible changes which can be made in this graph so that the audience does not miss this particular insight in the data.

Figure 6.2 provides a modified design of the same graph, with the aim of emphasizing the particular message of declining website sales.

The new design highlights the particular takeaway for which the graph was created. The graph uses five modified elements: a bold colour for the website line which stands out in contrast to the muted colours for the other channels, a thicker line for the website data, markers of a bigger size, an informative title of the same colour as the concerned category and a specific recommendation at the bottom of the chart.

These tactics of using a bold colour or a bigger marker for only one particular category trigger off what is called pre-attentive

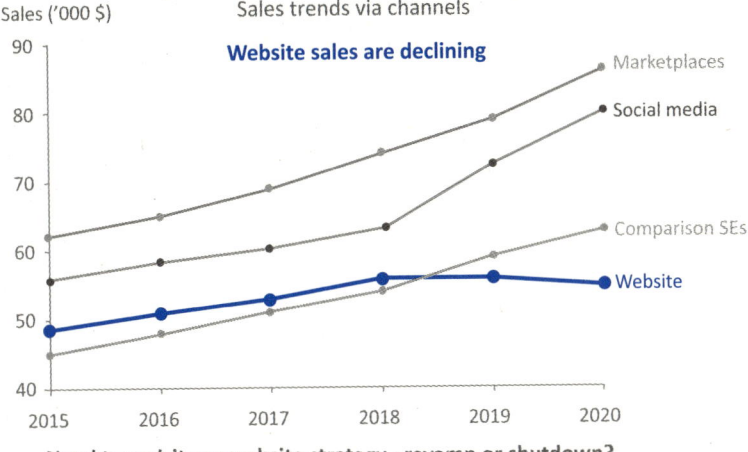

Figure 6.2: Pre-attentive attributes and textual elements have been used in this graphs to highlight a particular insight.

processing in the audiences' brain, forcing them to notice a particular part of the visualization.

PRE-ATTENTIVE PROCESSING AND HOW TO LEVERAGE IT

Pre-attentive processing refers to the unconscious and instantaneous processing that the human brain is capable of. This kind of data processing is different from the conscious or attentive processing which we normally attribute to the brain and which typically takes more time than pre-attentive processing. Pre-attentive processing is capable of recognizing a limited set of visual attributes which are appropriately named pre-attentive attributes. Our brains can instantaneously pick out objects which differ on any of these attributes, making them ideal for directing the viewer's attention to a particular aspect of the data.

Table 6.1: List of pre-attentive attributes useful for data visualization.

Colour	Hue
	Intensity
Form	Orientation
	Length and width
	Marks added to objects
	Shape
	Size
	Enclosure
Motion	Flicker
	Direction
Spatial Position	2D position
	Grouping

Colin Ware in his book on information design[1] identifies four broad categories of pre-attentive attributes: colour, form, motion and spatial position. Table 6.1 lists a subset of these pre-attentive attributes useful for visualizing data.

Attributes of colour: Objects of a different hue stand out immediately. Within objects of the same hue, those with a different intensity stand out more. In particular, a more intense shade of a particular hue draws the viewer's attention better.

O O O O O

O O O O O

Notice how one particular 'O' stands out in each row due to a different hue (top row) or a different intensity (bottom row).

Attributes of form: A change in orientation or slant catches the viewer's attention. For instance, using italics is a sure way to

---

[1] Colin Ware, *Information Visualization: Perception for Design* (2nd ed., San Francisco: Morgan Kaufmann, 2004).

highlight a specific word. This is particularly important to keep in mind if you feel tempted to use different fonts in a visualization. If you do not want the textual elements in your graph to draw attention away from the actual data elements, it is prudent to use a standard font throughout the visual. The example below illustrates the highlighting effect of using a different font type.

S   S   S   S   S

Longer or shorter objects stand out as do objects which are wider or thicker. Added marks on objects also make them more conspicuous. Objects with a different size or shape or those which are enclosed stand out more than their counterparts. Figure 6.3 contains examples of six preattentive attributes of length, width, added marks, shape, size and enclosure.

Let's revisit the line chart in Figure 6.2. Can you identify the three pre-attentive attributes used to draw the viewer's attention? The chart uses a different hue, a thicker line and bigger markers to distinguish one particular category.

In addition to hue, thickness (width) and size, other attributes which can be harnessed to direct user attention to a particular

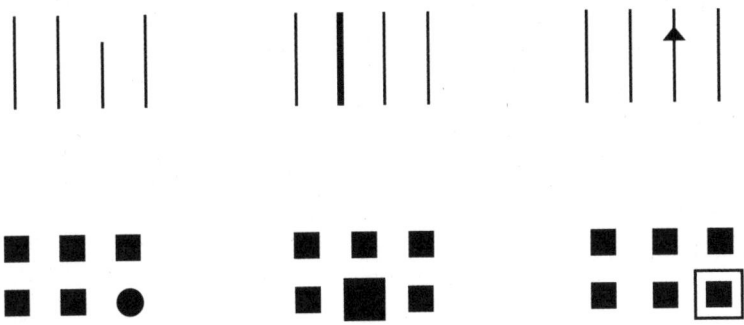

Figure 6.3: Examples of pre-attentive attributes of length, width, added marks, shape, size and enclosure. Notice how easily the odd object in each group draws your attention.

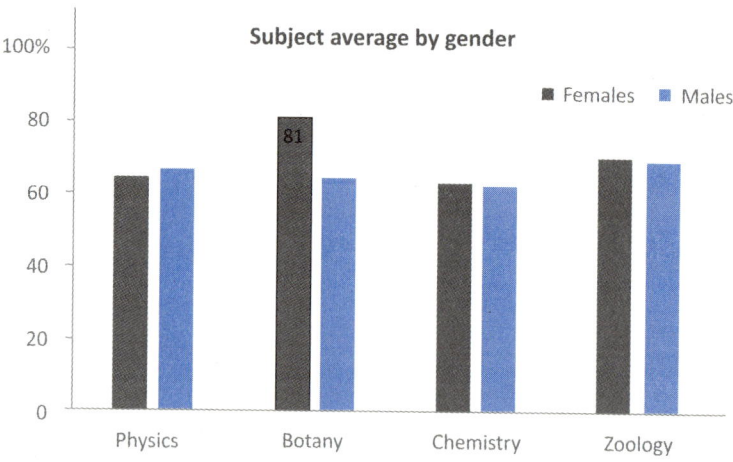

Figure 6.4: Pre-attentive attributes of enclosure (border) and added marks (label) help highlight the unusually high marks in botany for female students.

element in a graph include added marks and enclosure. The example in Figure 6.4 depicts a possible way to use these elements in a chart. The bar chart contains average marks across four subjects for students in an exam. The intention of this visualization is to draw attention to the unusually high scores for female students in botany. The pre-attentive attributes of enclosure (border around bar) and added marks (label) have been used to direct the viewer to this particular piece of information. Additionally, a more intense shade of grey could have been used to further highlight the particular bar but might be an overkill in this case, as the other two attributes are sufficient to highlight that particular data point.

Attributes of motion: Flickering objects or words immediately draw our attention. While a regular static graph doesn't use motion or animation, flicker is a great tool to use for real-time data dashboards. Suppose a critical metric which is being monitored regularly on a dashboard falls dangerously below its

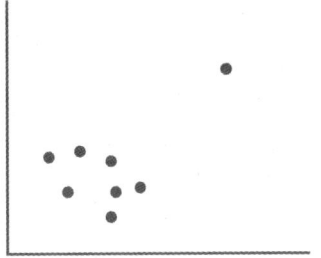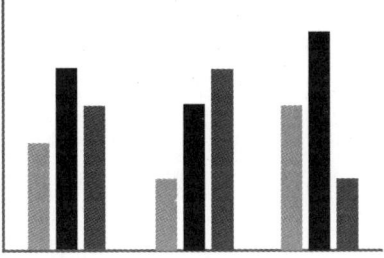

Figure 6.5: Pre-attentive attribute of proximity helps identify the outlier in the scatterplot (left) and groups bars into three categories in the bar graph (right).

threshold. Flickering icons or messages on the dashboard can draw the attention of the user.

Animations or flying objects/words are sometimes used in presentations to catch the audience's attention. While moving objects do force us to notice them, they are best reserved as a ploy to wake up a sleepy audience. Too many moving objects can bewilder your audience and leave them dizzy!

Motion charts like the animated scatterplots created by Gapminder[2] are catchy and useful in a limited set of scenarios but mostly not very useful for analysing data.

Attributes of spatial position: Objects positioned together tend to be seen as a group and objects spatially distant from a cluster tend to stand out. This pre-attentive attribute helps us easily identify exceptions—think of a scatterplot with outliers. Another application of spatial positioning is in clustered bar graphs, where bars bunched close together are understood as belonging to the same category. Both these applications of proximity are illustrated in Figure 6.5.

As the above examples illustrate, pre-attentive attributes are effective instruments to not only organize your data but also

---

[2] www.gapminder.org/tools

help direct the audience to a particular part of the visualization. Use one or more of these attributes to bring out the intended insight in your graph and tell your data story.

Apart from pre-attentive attributes, textual elements like a title or annotation can play an important role in articulating the intended message of a data visualization. The example in Figure 6.2 on sales trends via different channels uses two textual elements to drive the narrative, a title at the top of the graph and a call to action at the bottom. The effective use of these and other textual elements are discussed in the next subsection.

THE ROLE OF TEXT IN A DATA VISUAL

Textual elements in graphs have been singled out in data visualization literature as important components which can be leveraged to enhance the message in the data. They have on occasion been hailed as the primary storytelling element!

> Textual annotations are not mere sidekicks that assist data visualizations that convey information; they can be the storyteller with the visualization there to back up the credibility of its message.[3]

Stephen Few in his book *Show Me the Numbers* identifies eight possible applications of text in a graph: to label categories, introduce the visual, explain the insight, reinforce the main message, highlight certain parts of the visual, create a sequence

---

[3] Ha-Kyung Kong, Zhicheng Liu, and Karrie Karahalios, 'Trust and Recall of Information across Varying Degrees of Title—Visualization Misalignment', *CHI '19: Proceedings of the 2019 CHI Conference on Human Factors in Computing Systems*, May 2019, https://doi.org/10.1145/3290605.3300576

if multiple graphs are contained in the same view, recommend a course of action or propose possible directions for further enquiry.

The title of a data visualization is an important textual element which can play a crucial role in either introducing the visual or explaining the message. There are mainly two different types of titles to choose from: a generic or basic title which states what the visual contains or an informative title (sometimes called an action title or descriptive title) which explains the main message or 'takeaway' in the visual.

Revisiting the example in Figure 6.4, the title states 'Subject Average by Gender'. This is a generic title as it doesn't draw attention to the specific insight in the data. An example of an informative title for this graph could be as follows: 'Female students vastly outperformed male students in botany.'

This title straightaway cues the user to the intention of the graph and ensures that the message is not missed. Some visualization experts recommend using a title and a subtitle[4]—a generic short title followed by a longer descriptive subtitle.

In a study which specifically tries to understand what makes a visualization effective, Borkin and co-authors[5] measured which elements of a graph users look at most and store in their memory and also what they could recall about the visual. The study found that the title and other textual elements of the graph were the most recalled elements. Interestingly, the study

---

[4] http://stephanieevergreen.com/wp-content/uploads/2017/03/DataViz Checklist_May2016.pdf

[5] Michelle A. Borkin, Zoya Bylinskii, Nam Wook Kim, Constance May Bainbridge, Chelsea S. Yeh, Daniel Borkin, Hanspeter Pfister, and Aude Oliva, 'Beyond Memorability: Visualization Recognition and Recall', *IEEE Transactions on Visualization and Computer Graphics* 22 (January 2016): 519–528.

concluded that a well-thought-out title helps viewers better understand and remember the main message of the visualization.

The authors also found that redundancy (in other words, repeating the takeaway in more than one way) helped with better recall and understanding of the visualization. Their finding points to the usefulness of reinforcing the message in the data with elements like an informative title or a concluding statement at the bottom of the visual.

Another study[6] which explicitly tests the impact of different types of titles on the effectiveness of a visualization found that informative titles required less mental effort when compared to generic titles. The study also found that informative titles were more useful when used with complex graphs rather than simple ones. The results from both these studies point to the potential of textual elements in helping a viewer understand and remember the main message in a graph.

Revisiting our earlier example of gender-wise performance in science subjects, the first version in Figure 6.4 used pre-attentive attributes of enclosure and added marks to draw the attention of the audience to a particular part of the graph. The revised version in Figure 6.6 uses an informative title and colour intensity to highlight the intended message. Which version works better for you in terms of understanding the purpose of the graph?

While the first version sufficiently highlights the bar representing the performance of female students in botany, the intended message of the visualization is easier to grasp in the second version, owing to the action title.

---

[6] Dana Linnell Wanzer, Tarek Azzam, Natalie D. Jones, and Darrel Skousen, 'The Role of Titles in Enhancing Data Visualization', *Evaluation and Program Planning* 84, no. 3 (February 2021): 101896.

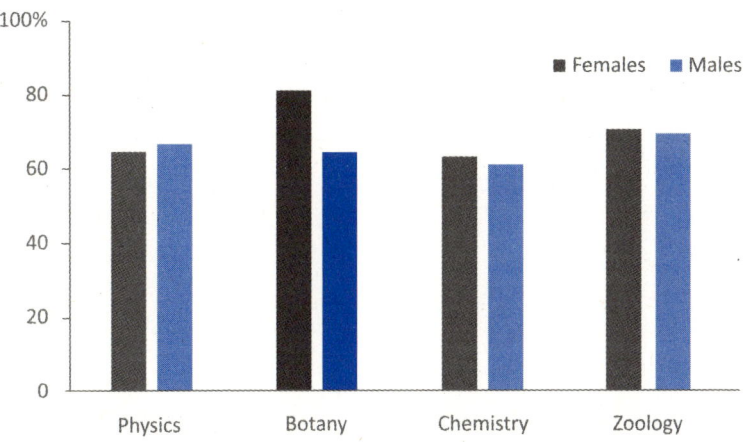

Female students vastly outperformed male students in Botany

Figure 6.6: Informative titles (sometimes called action titles) can play an important role in communicating the key message in a visualization.

Do note that relevant portions of the action title in Figure 6.6 are coloured the same as the respective bars—dark grey for botany scores for females and dark blue for botany scores for males. Using such a colour coding makes it easier for the viewer to link the message in the action title to the relevant parts of the visualization. The particular colour used for such a purpose is popularly called the 'action colour'.

Another textual element which can vastly improve the understanding of a graph is an annotation. While an informative title states the message in the data, it is placed relatively far away from where the data is drawn. This sometimes becomes problematic, especially if there are multiple focus points in a graph. Annotations, on the other hand, allow the textual explanations to be placed near the relevant visual explanations, saving many back and forth eye movements for the viewer. The power of well-placed annotations can clearly be seen in the example in Figure 6.7, which depicts unemployment rates in the USA over a span of 17 years.

U.S. unemployment rate

Unemployment triggered by Covid reached almost 15%, surpassing the rate of around 10% during the Great Depression of 2007–2009

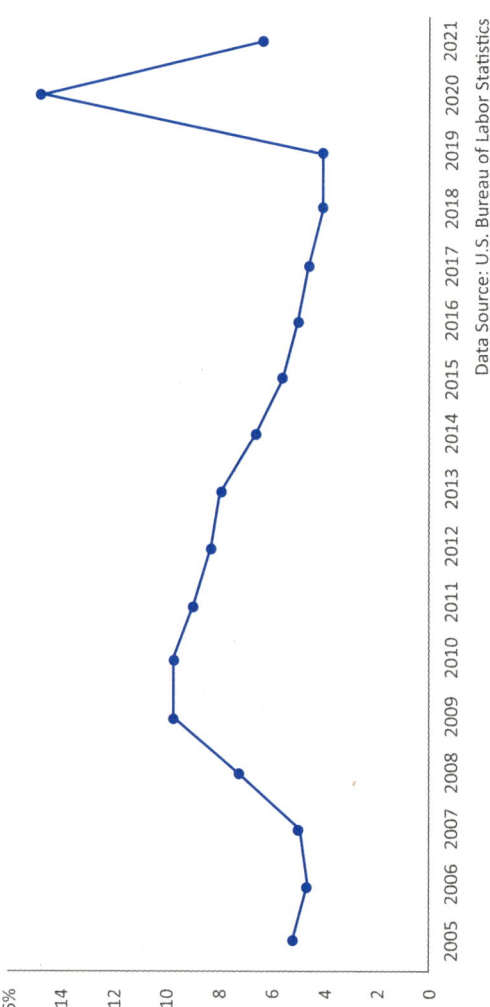

Data Source: U.S. Bureau of Labor Statistics

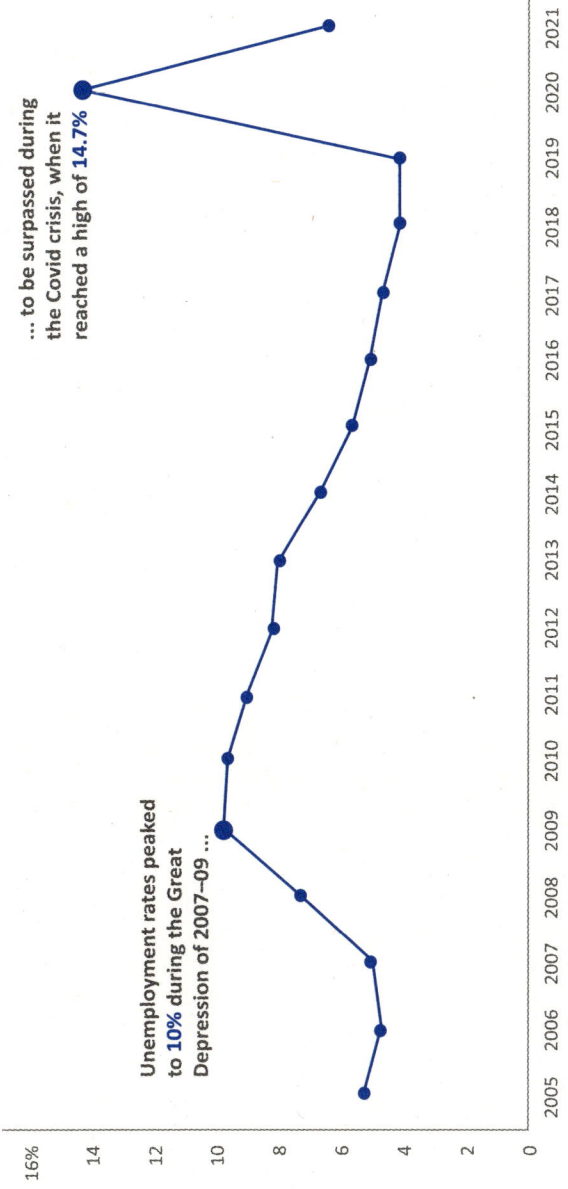

Figure 6.7: The graph on top uses an informative title which forces viewers to hunt for the relevant parts of data, whereas the bottom uses annotation to directly focus on the pertinent portions of the line.

Data source: US Bureau of Labor Statistics.

While the first version uses an informative title to spell out the message, the second version with annotations is far more effective in helping the viewer focus on the relevant parts of the line graph. Indeed, a recent empirical study[7] aimed at testing the usefulness of these design guidelines found that adding focus to a graph (using annotations and/or colour) led to almost a three times higher recall of the pattern in the data. Additionally, the test subjects rated the focused graphs as more aesthetic and having more clarity.

Pre-attentive attributes and textual elements are powerful tools at your disposal to help create a data visual which speaks your message loud and clear. However, a note of caution to the reader: The power of titles and annotations is a double-edged sword. Misleading titles can quite easily bias the understanding of the data as well. An experimental study[8] which specifically looked at the impact of slanted titles in visualizations found that depending on how the title was worded, viewers derived completely opposite messages from the same visualization.

Much like the sleight of hand used by magicians to distract the audience, pre-attentive attributes combined with misleading text have been used to misdirect the viewers' attention and lead them to incorrect understandings of the data.

---

[7] K. Ajani, E. Lee, C. Xiong, C. Nussbaumer Knaflic, W. Kemper, and S. Franconeri, 'Declutter and Focus: Empirically Evaluating Design Guidelines for Effective Data Communication', *IEEE Transactions on Visualization and Computer Graphics*, https://visualthinking.psych.northwestern.edu/publications/Ajani_Declutter_2021.pdf

[8] Ha-Kyung Kong, Zhicheng Liu, and Karrie Karahalios, 'Frames and Slants in Titles of Visualizations on Controversial Topics', *CHI '18: Proceedings of the 2018 CHI Conference on Human Factors in Computing Systems*, April 2018, https://doi.org/10.1145/3173574.3174012

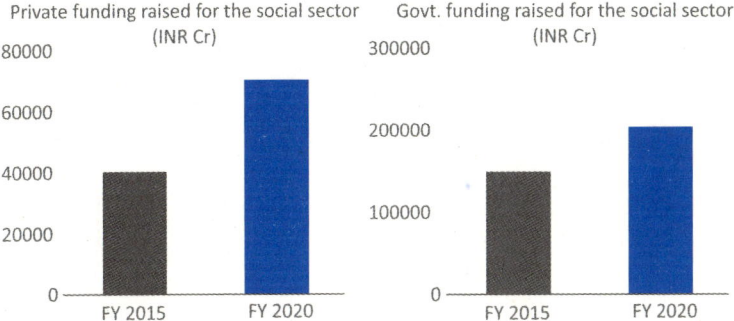

Private funding grew at a 15% annual rate, surpassing public funding's growth at 10%

Figure 6.8: This visualization uses the pre-attentive attribute of hue to encourage comparison of bars which don't share a common scale.

Figure 6.8 illustrates this by example.[9]

The two graphs in the example above show funding raised for the social sector from private and public sources, for the years 2015 and 2020. The title in blue and the two blue bars stand out among the other elements. By leveraging the pre-attentive attribute of hue, the viewer is encouraged to compare the two blue bars. This, along with the suggestion in the title, could lead the viewer to conclude that private funding has increased much more than government funding.

But look again—is that really true? Did you notice that the y-axis scale is very different for the two graphs, rendering the comparison of the lengths of the blue bars meaningless? Even though private funding has increased by 15 per cent versus 10 per cent for government funds, in absolute figures, private funding increased by only ₹30,000 crore compared to ₹65,000 crore from government sources.

---

[9] A similar graph was used in the *India Philanthropy Report 2019*, available at www.bain.com/insights/india-philanthropy-report-2019

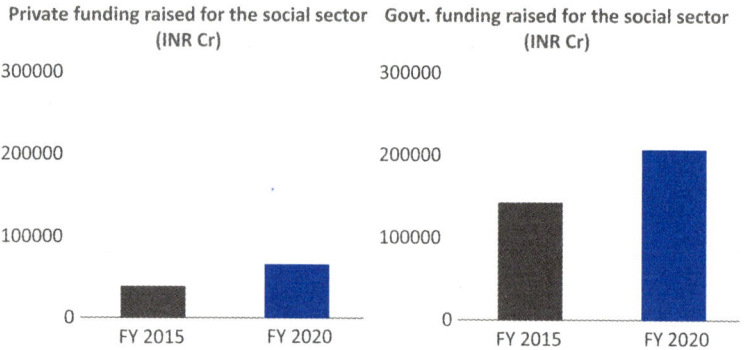

Private funding is still only a third of public funding

Figure 6.9: The corrected version uses the same scale for both graphs, facilitating a meaningful comparison across the two graphs.

In the corrected visualization in Figure 6.9, both graphs have the same quantitative scale, making it easy to see that private funding is still only a fraction of government funding.

Beware of graphs which deceive with misleading titles and use pre-attentive attributes to distract the viewer from the truthful message in the data.

PUTTING IT TOGETHER: WEAVING A NARRATIVE

Stories help the audience empathise better with whatever is being told and are known to have a higher recall value. Telling an effective story with data involves many different elements, some of which we have already covered so far. The five-pronged approach[10] discussed below provides a systematic framework

---

[10] For a detailed discussion, refer to Kavitha Ranganathan, 'Creating Compelling Stories with Data: The Five-Pronged Approach', Technical Note (Indian Institute of Management, Ahmedabad, 2020), https://cases.iima.ac.in/index.php/creating-compelling-stories-with-data-the-five-pronged-approach.html

for putting together these different elements and creating a persuasive data story. The five steps include knowing your audience, articulating the big idea, constructing a narrative arc, identifying the best visual construct for the situation and leveraging additional storytelling elements. Each step in the five-pronged approach is discussed in detail below.

Know Your Audience

Identify who the key decision-makers are in your audience. What knowledge does your audience already have and what context should they be made aware of? What jargon or terms might be familiar or unfamiliar to your audience? What matters to them and why should they care about what you are presenting? What visual constructs would be familiar or unfamiliar to them? What colour themes might they prefer? Pay particular attention while choosing colours to represent particular insights, as different cultures associate different emotions with particular colours. There are many helpful sources online[11] to understand how different colours are interpreted across different cultures.

Articulate Your Big Idea

Clearly identify the purpose of your exercise. What is the message you want your audience to go home with? Try to articulate the big idea of your project by creating a complete sentence that captures your point of view regarding what the data is revealing, what is at stake and your recommendation for future action. The big idea does not necessarily need to be presented to the

---

[11] https://summalinguae.com/language-culture/colours-across-cultures/

audience directly; it is a behind-the-scenes preparation that you do to help clarify your objective and organize your thoughts.

Construct the Narrative Arc

You can leverage the classic three-act story structure (also known as Freytag's pyramid) to demarcate the narrative flow into three sections. The first section should usually set the context and introduce the current situation. The second section can discuss the problem or conflict and the third section of your presentation should explain the proposed solution or recommendation. In some situations, an alternate flow can be considered, where the third act (the recommendation) comes first, and then the context and finally the detailed explanation of the problem. Consultants often use this alternate construct in their presentations.

Identify the Best Visual Construct (Graph)

As a first step, identify the places in your narrative where data evidence needs to be presented. For each of these situations, identify what relationship in the data needs to be conveyed. As discussed in the first chapter of the book, some common data relationships are time series, ranking, part-to-whole, distribution and deviation.

Certain situations could involve a combination of these relationships as well. Part-to-whole and time series is a classical combination, especially when we need to know changes in the part-to-whole relationship over time. For example, you may want to present the market share for your company and your competitors and how these percentages have changed over time. Lines or a series of stacked bar graphs (especially if there are three or less categories) could work well for such a scenario.

The guidelines for choosing the best visual constructs for each of these relationship have been discussed in detail in the beginning of the book.

Leverage Additional Storytelling Elements

As discussed earlier in the chapter, pre-attentive attributes such as hue, intensity, marker size or thickness of lines can be used to draw the audience's attention to the relevant parts of the graph. Textual elements such as action titles, annotations and recommendations are handy for clearly articulating the problem, a proposed solution or a call for action.

Animation can be a handy tool to build the visualization in layers. It is recommended that the axes be shown first so that the viewer can understand the layout and organization of the chart, then the data (bars, lines, etc.) should be shown and finally the additional textual elements which highlight the message. Animation can also be used to chronologically depict data, as this adds suspense to the narrative and builds the interest of the audience. Think about a line which extends from left to right (with time on the x-axis) accompanied by explanations from the presenter about the various peaks and drops it follows.

The five-pronged approach will help you gather and organize your thoughts and put together a compelling presentation or report which convinces your audience.

CLOSING THOUGHTS

There is virtually unlimited freedom in how we represent data. The difficult question is how best to represent it.[12]

This book has hopefully shown you that while there are many different ways to visually represent data, some are definitely better than others. Poorly designed visualizations can be ineffective in more than one way. At best, they may cause bewilderment or confusion and at worst the audience might believe something which is untrue. By alerting you to the myriad ways in which data can be misrepresented or misinterpreted, this book has tried to explain the art and science involved in creating accurate data visualizations.

While we cannot claim that there is only one correct way to visually represent a particular dataset, the art of choosing an effective and honest construct is much like the art of 'closing doors'.[13] While making a strategic decision, typically here are a number of options available or in other words a number of open doors we could potentially walk through. Choosing the right door to walk through entails knowing which other doors to close or, in other words, which options to disregard. This book will hopefully help you recognize and dismiss the erroneous options, guiding you to robust and intuitive designs instead, ultimately enabling you to create more accurate, intuitive and honest visualizations.

Now it is time to apply your learnings. The next section contains various exercises to practise your new-found knowledge on creating truthful and effective data visualizations.

---

[12] William Wright, 'Information Animation Applications in the Capital Markets', in *Readings in Information Visualization: Using Vision to Think* (San Francisco: Morgan Kaufmann Publishers, 1999).

[13] 'Strategy is the art of closing doors,' quote from Saral Mukherjee, *Elephants and Cheetahs: The Beauty of Operations* (New Delhi: Penguin, 2021).

PRACTISE YOUR LEARNINGS

Spot the problem with each of these graphs and suggest an honest redesign.

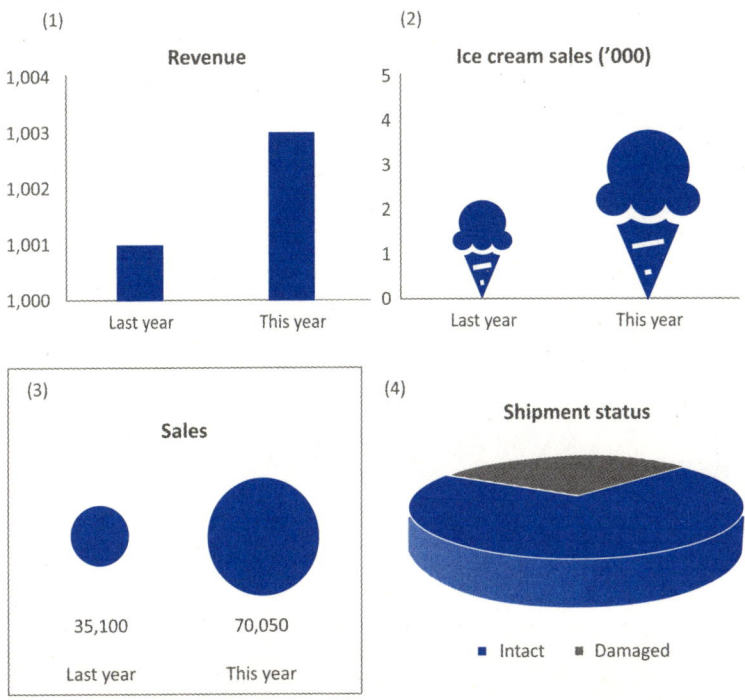

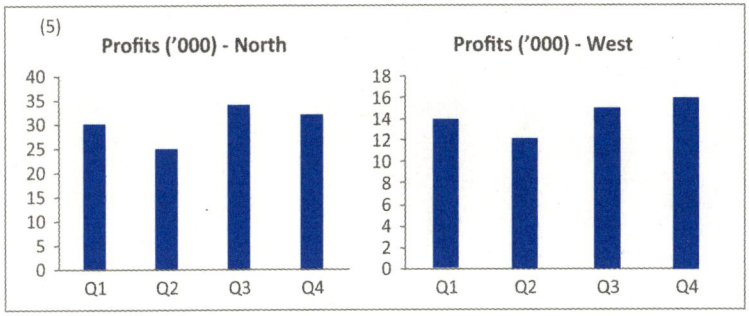

(6)

($) **Profits over time**

3,020
3,000
2,980
2,960
2,940
2,920
2,900

Oct   Sep   Aug   July

(7)

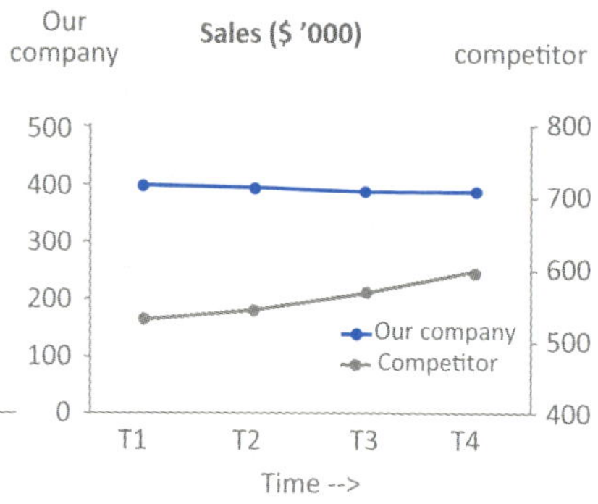

Our company   **Sales ($ '000)**   competitor

500 — 800
400 — 700
300 —
200 — 600
100 — 500
0 — 400

T1   T2   T3   T4

Our company
Competitor

Time -->

(8)   **Profits over time**

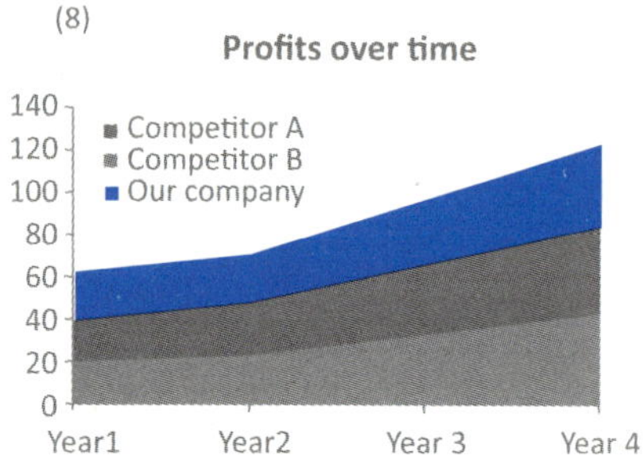

140
120
100
80
60
40
20
0

■ Competitor A
■ Competitor B
■ Our company

Year1   Year2   Year 3   Year 4

(9)   **Injuries on site**

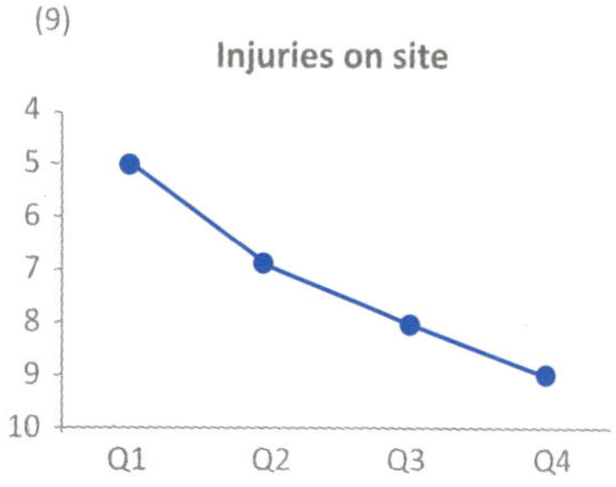

4
5
6
7
8
9
10

Q1   Q2   Q3   Q4

(10)

Number of employees   **Bonus payouts**

70
60
50
40
30
20
10
8

0–4%   5–9%   10–14%   15–25%

% of annual salary

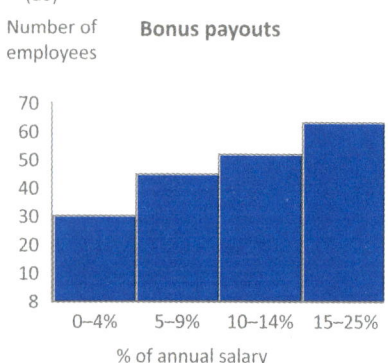

(11)

Profits ($'00)   **Discounts vs profits (Quarter1 - Quarter4)**

18
16
14
12
10
8

15   20   25   30   35   40

Discount (%)

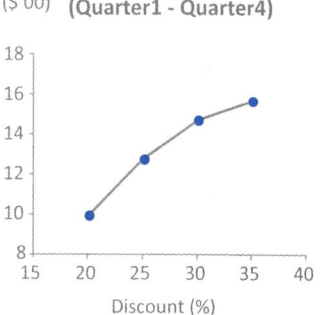

Solutions for Practice Problems

(1) Revenue: Bar Chart

The y-axis scale does not start at zero; hence, it has highly exaggerated the growth in revenue. The corrected bar chart has a scale starting at zero.

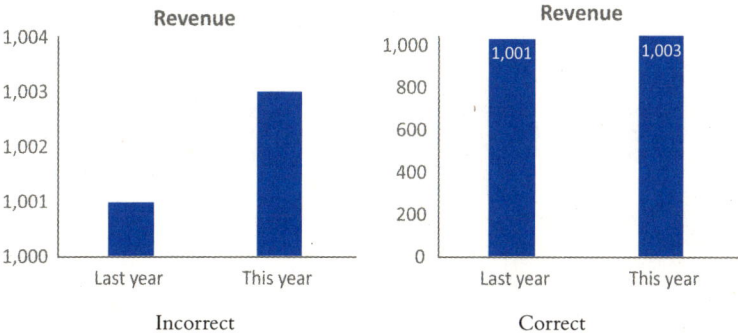

Incorrect   Correct

(2) Ice Cream Sales: Icons

Although sales have only doubled, the icon has doubled in both height and width, giving an exaggerated impression of increase in ice cream sales. The accurate representation uses multiple icons of the same size placed one on top of the other.

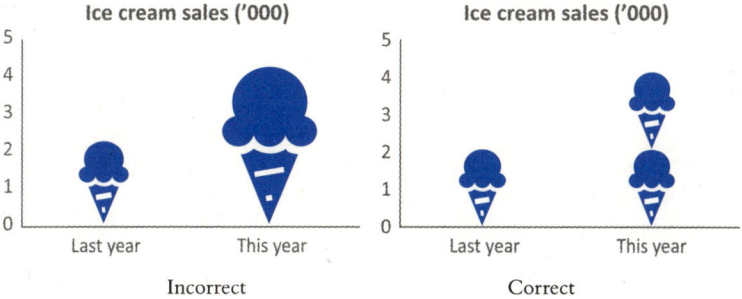

Incorrect   Correct

(3) Sales: Circles

The quantity is encoded in the diameter of the circles instead of the area, again exaggerating the growth in sales. In the redesigned version, the quantities are encoded as areas of the two circles.

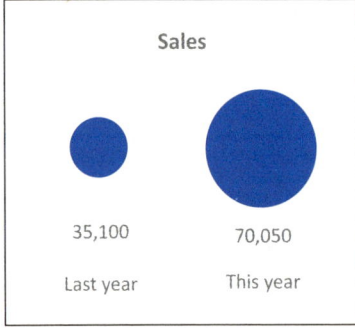

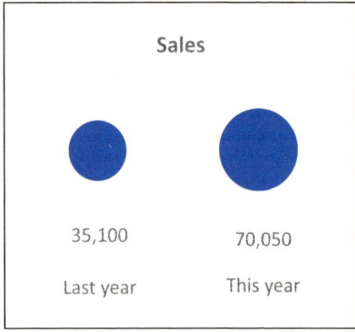

Incorrect   Correct

(4) Shipment Status: 3D Pie

Nearly one out of three consignments were damaged, but by placing the 'damaged' pie at the back of the 3D pie, the damages look much smaller. A 2D pie chart is more useful to judge the actual ratios of the two categories.

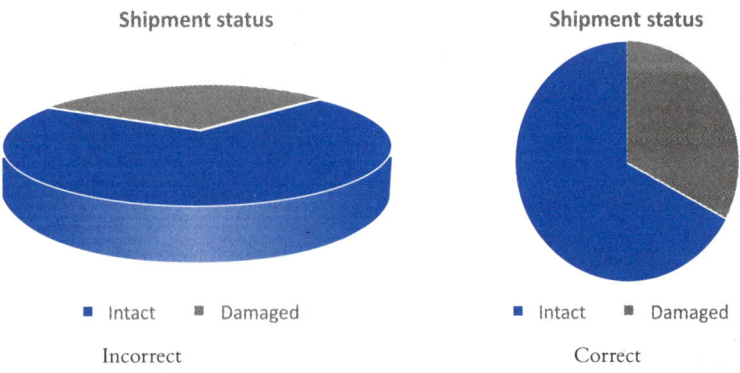

Incorrect   Correct

(5) Zone-Wise Profits: Small Multiples

Inconsistent y-axis scales in the two graphs make comparisons across the two graphs meaningless. Profits in the west zone seem to match those in the north but are actually much lesser. Both scales should be exactly the same for a small multiples construct.

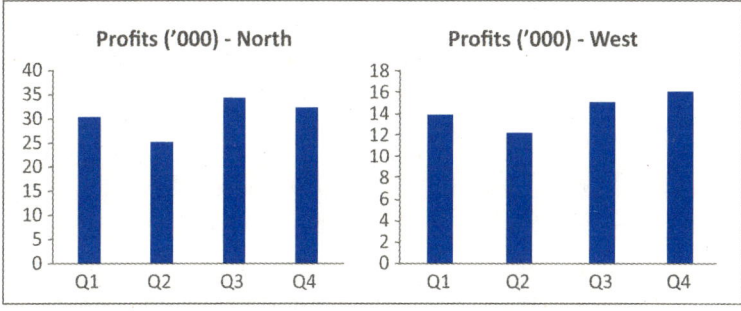

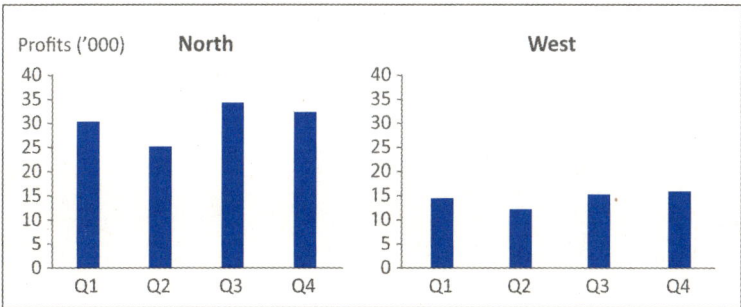

Incorrect (top)
Correct (bottom)

(6) Profits over Time: Trend Line

Monthly profits seem to have increased but have actually decreased. The months are arranged from right to left. Viewers generally expect time to flow from left to right.

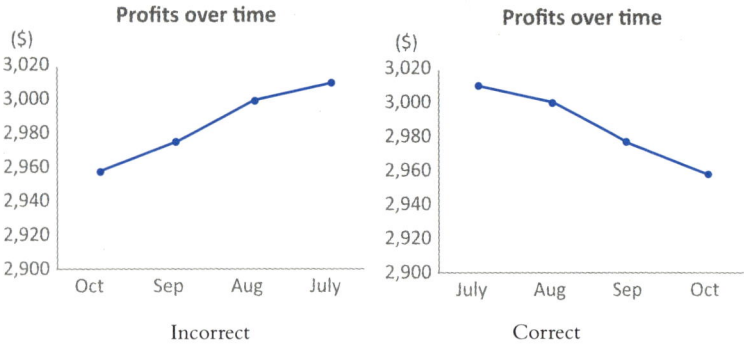

(7) Sales: Dual Axis Graph

Sales for 'Our Company' are much lower than the competitor's, but the line is visually higher because of two separate scales, creating confusion. Using only one scale (since the units are the same for both variables) allows accurate comparison of sales of the two companies.

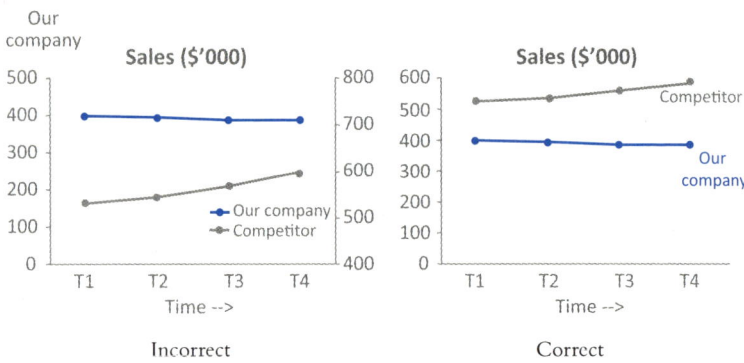

(8) Profits over Time: Stacked Area Chart

'Our Company's' profits seem to increase at a faster rate than the competition. But the increase is exaggerated because the blue band is placed on a baseline which is also curved upwards. The trend line we see is actually the cumulative profits for all three companies. Individual (non-stacked) lines for each company reveal that the competitors have actually done better in terms of profits.

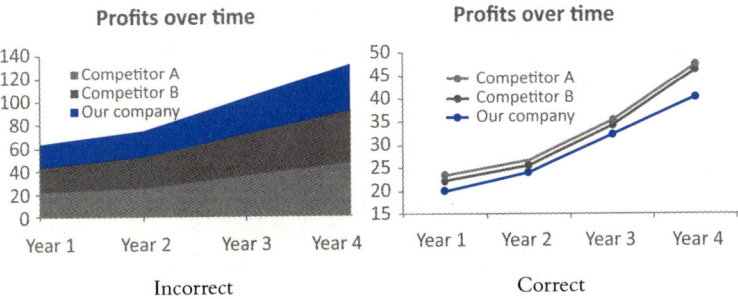

(9) Injuries on Site: Line Graph

The number of injuries per quarter seems to be decreasing but only because the y-axis scale is reversed. Except for the case where the scale represents ranks, most viewers will expect the scale to start from the bottom.

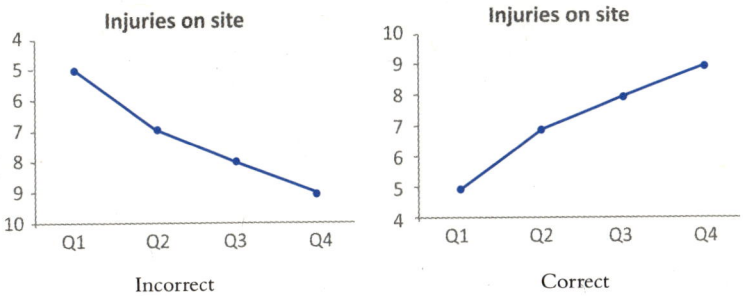

(10) Bonus Payouts: Histogram

The last bar has a larger bin width than the others. This leads to the wrong impression that the highest bonus category has the most employees. An accurate histogram as shown in the redesign should have bins of uniform width (range).

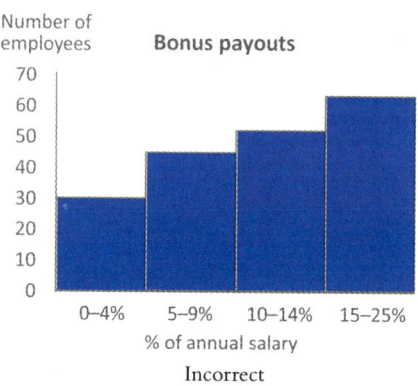

Incorrect

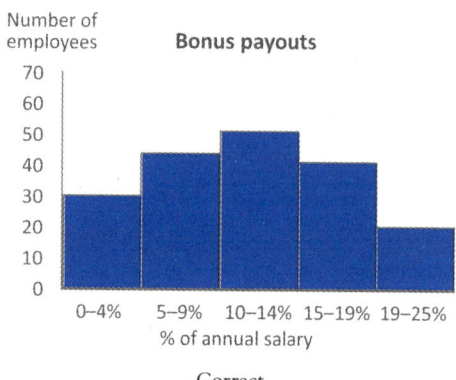

Correct

(11) Discounts versus Profits: Connected Scatterplot

With no arrows or any other indication of the chronology of the dots, the viewer might assume that time is flowing from left to right. In this case, the viewer might incorrectly assume that both discounts and profits have increased across the quarters. Arrows or another markings should clearly indicate the direction of time in a connected scatterplot.

ACKNOWLEDGEMENTS

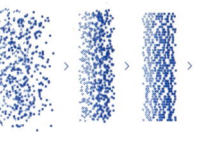

It takes a village to raise a child and it took a network of family and friends to write this book. I would first like to thank the folks at Penguin Random House India, especially Radhika and Ralph for their committed presence through the whole process. Additionally, a big thank you to the team at Sage—Manisha, Neena and Sandhya—for their support for the earlier edition of the book. I am also grateful to all the participants in my classroom, across the years. Your enthusiasm for the subject and encouraging feedback (sometimes years later) helped create and shape the book.

This book was chiefly written during the pandemic, with two kids at home trying to attend online school and two adults trying to work and an ebullient puppy trying very hard to contribute by jumping on the laptop's keypad repeatedly. The manuscript was my safe space and anchor that I could go back to repeatedly, when the outside world seemed to be crumbling around us. There were numerous occasions when the writing came to a standstill due to unforeseen emergencies unfolding and I have Akshaya, Divya and Pritha to thank for their gentle prodding about the progress, which would guilt me into jumpstarting the writing again.

My loving extended family in Bangalore, Jamshedpur and Singapore helped with so much enthusiasm whenever I needed their inputs. I owe a lot to all of them, including little Zoya, the youngest, who cheers us all up with her radiant bounciness. Papa and Mummy were steadfast supporters of this project, having gone through book projects of their own and knowing the grind involved to see one through. The first four graphs you see in this book were tested on my mother and the idea for the title came from my sister (along with a host of other very innovative title ideas from both, it's a pity I had to choose only one). Thank you both for your shining presence. My dad has been a rock through all the ups and downs in my life and I owe my persistence and stubborn resilience to him.

Thank you, Adya and Avital for your patience. Your mom was often lost in formulating passages for the book and would stare blankly at you when you tried to get her attention. Finally, and most importantly, thank you Ankur, for your steadfast support and love through the whole process. I owe my adventures in data visualization and much more to you.